**"十四五"职业教育国家规划教材**

中等职业教育专业技能课教材
中等职业教育中餐烹饪专业系列教材

# 中餐烹饪英语

ZHONGCAN PENGREN YINGYU（第2版）

主　编　张艳红　赵　静
副主编　向　军　牛京刚　刘欣雨
参　编　钱　俊　姜　珊　夏　孟　李　蕊
　　　　葛苗苗　宋　洋　孙　敏

重庆大学出版社

## 内容提要

本书分为10个单元。每个单元编写体例一致，由You will be able to, Warming Up, Now I Can Learn, Now I Can Speak, Now I Can Read, Now I Can Do, Word List, Learning Tips 和Culture Knowledge 9部分组成。本书以工作任务为导向，采用情景模拟、角色扮演等多种方法学习单词、词组和对话，呈现形式以听、读、说为主，引导学生在掌握单词的基础上模拟真实的工作场景和内容进行对话练习，同时进行阶段性自我学习评价，增强学生学习的信心。

本书还配有电子教学资源，包括习题答案、录音教学资料、电子课件等。学生扫码即可收听单词、词组、对话等，方便学习。本书可作为中等职业教育中餐烹饪专业教材，也可作为广大烹饪爱好者的参考用书。

**图书在版编目（CIP）数据**

中餐烹饪英语 / 张艳红，赵静主编. --2版. -- 重庆：重庆大学出版社，2023.10
中等职业教育中餐烹饪专业系列教材
ISBN 978-7-5624-8754-8

Ⅰ.①中… Ⅱ.①张…②赵… Ⅲ.①中式菜肴—烹饪—英语—中等专业学校—教材 Ⅳ.①TS972.117

中国版本图书馆 CIP 数据核字(2021)第263657号

中等职业教育中餐烹饪专业系列教材
**中餐烹饪英语**
（第2版）

主　编　张艳红　赵　静
策划编辑：沈　静

责任编辑：陈亚莉　沈　静　　版式设计：沈　静
责任校对：谢　芳　　　　　　　责任印制：张　策

＊

重庆大学出版社出版发行
出版人：陈晓阳
社址：重庆市沙坪坝区大学城西路21号
邮编：401331
电话：（023）88617190　88617185（中小学）
传真：（023）88617186　88617166
网址：http://www.cqup.com.cn
邮箱：fxk@cqup.com.cn（营销中心）
全国新华书店经销
重庆长虹印务有限公司印刷

＊

开本：787mm×1092mm　1/16　印张：9.5　字数：296千
2015年11月第1版　2023年10月第2版　2023年10月第4次印刷
印数：8 001—10 000
ISBN 978-7-5624-8754-8　定价：45.00元

本书如有印刷、装订等质量问题，本社负责调换
版权所有，请勿擅自翻印和用本书
制作各类出版物及配套用书，违者必究

## 中等职业教育中餐烹饪专业系列教材
### 主要编写学校

北京市劲松职业高中

北京市外事学校

上海市商贸旅游学校

上海市第二轻工业学校

广州市旅游商务职业学校

江苏旅游职业学院

扬州大学旅游烹饪学院

食品科学与工程学院

河北师范大学旅游学院

青岛烹饪职业学校

海南省商业学校

宁波市古林职业高级中学

云南省通海县职业高级中学

安徽省徽州学校

重庆市旅游学校

重庆商务职业学院

## 出版说明

2012年3月19日，教育部职业教育与成人教育司印发《关于开展中等职业教育专业技能课教材选题立项工作的通知》（教职成司函〔2012〕35号），我社高度重视，根据通知精神认真组织申报，与全国40余家职教教材出版基地和有关行业出版社积极竞争。同年6月18日，教育部职业教育与成人教育司致函（教职成司函〔2012〕95号）重庆大学出版社，批准重庆大学出版社立项建设中餐烹饪专业中等职业教育专业技能课教材。这一选题获批立项后，作为国家一级出版社和教育部职教教材出版基地的重庆大学出版社珍惜机会，统筹协调，主动对接全国餐饮职业教育教学指导委员会，在编写学校邀请、主编遴选、编写创新等环节认真策划，投入大量精力，扎实有序推进各项工作。

在全国餐饮职业教育教学指导委员会的大力支持和指导下，我社面向全国邀请了中等职业学校中餐烹饪专业教学标准起草专家、全国餐饮职业教育教学指导委员会委员和委员所在学校的烹饪专家学者、一线骨干教师，以及餐饮企业专业人士，于2013年12月在重庆召开了"中等职业教育中餐烹饪专业立项教材编写会议"，来自全国15所学校30多名校领导、全国餐饮职业教育教学指导委员会委员、专业主任和一线骨干教师参加了会议。会议依据《中等职业学校中餐烹饪专业教学标准》，商讨确定了25种立项教材的书名、主编人选、编写体例、样章、编写要求，以及配套电子教学资源制作等一系列事宜，启动了书稿的撰写工作。

2014年4月，为解决立项教材各书编写内容交叉重复、编写体例不规范统一、编写理念偏差等问题，以及为保证本套立项教材的编写质量，我社在北京组织召开了"中等职业教育中餐烹饪专业立项教材审定会议"。会议邀请了时任全国餐饮职业教育教学指导委员会秘书长桑建先生、扬州大学旅游烹饪学院路新国教授、北京联合大学旅游学院副院长王美萍教授和北京外事学校高级教师邓柏庚组成审稿专家组对各本教材编写大纲和初稿进行了认真审定，对内容交叉重复的教材在编写内容划分、表述侧重点等方面作了明确界定，要求各门课程教材的知识内容及教学课时，要依据全国餐饮职业教育教学指导委员会研制、教育部审定的《中等职业学校中餐烹饪专业教学标准》严格执行，配套各本教材的电子教学资源坚持原创、尽量丰富，以便学校师生使用。

本套立项教材的书稿按出版计划陆续交到出版社后，我社随即安排精干力量对书稿的编辑加工、三审三校、排版印制等环节严格把关，精心安排，以保证教材的出版质量。本套立项教材第1版于2015年5月陆续出版发行，受到了全国广大职业院校师生的广泛欢迎及积极选用，产生了较好的社会影响。

在本套立项教材大部分使用4年多的基础上，为适应新时代要求，紧跟烹饪行业发展趋势和人才需求，及时将产业发展的新技术、新工艺、新规范纳入教材内容，经出版社认真研究于2020年3月整体启动了本套教材的第2版全新修订工作。第2版修订结合学校教材使用反馈情况，在立德树人、课程思政、中职教育类型特点，以及教材的校企"双元"合作开发、新形态立体化、新型活页式、工作手册式、1+X书证融通等方面做出积极探索实践，并始终坚持质量第一，内容原创优先，不断增强教材的适应性和先进性。

在本套教材的策划组织、立项申请、编写协调、修订再版等过程中，得到教育部职业教育与成人教育司的信任、全国餐饮职业教育教学指导委员会的指导，还得到众多餐饮烹饪专家、各参编学校领导和老师们的大力支持，在此一并表示衷心感谢！我们相信本套立项教材的全新修订再版会继续得到全国中职学校烹饪专业师生的广泛欢迎，也诚恳希望各位读者多提改进意见，以便我们在今后继续修订完善。

重庆大学出版社<br>
2021年7月

# 前言

（第 2 版）

随着我国经济的不断发展，尤其是在北京成功举办了 2008 年奥运会之后，世界各国的游客不断涌入中国，极大地促进了旅游业的发展。随着旅游业的高速发展，对餐饮行业从业者的英语水平提出了更高的要求。本书涵盖厨房中各个加工间的专业英语知识，紧密联系厨房岗位工作实际，训练学生基本的英语听、说、读、写能力。通过学习，学生能获取从事烹饪行业岗位所需英语交际的实际技能和知识，为学生未来的职业发展和专业学习打下坚实的语言基础。本书是中等职业教育中餐烹饪专业英语课程用书，本书也可以作为旅行社培训、烹饪从业人员和中餐美食爱好者的自学教材。2023 年，本书被评为"十四五"职业教育国家规划教材。目前，本书已经多次印刷。

本书具有以下特点。

1. 图文并茂，易学易记

本书以学生为中心，配备了大量彩色图片，图文并茂，浅显直观，易学易记。设计的生词、词组、对话内容，能够紧密联系厨房岗位工作实际，达到激发学生学习兴趣的目的。

2. 构思新颖，实践性强

本书构思新颖，以一名中餐烹饪专业实习生 Jack 到饭店顶岗实习过程中的所见、所学、所感为线索，以情景教学为基石，以活动为内容，强调语言学习的实践性。本书将英语教学与专业教学有机结合，既可作为英语教材，又可辅助专业教师实行"双语教学"。

3. 模块设置独特，学习方式灵活自由

每个单元均设置了 Learning Tips 和 Culture Knowledge 两个模块。在 Learning Tips 中补充了与单元内容相关的烹饪专业知识，在 Culture Knowledge 中丰富了餐饮背后的文化生活，扩宽了学生的视野，让学生树立大食物观。同时，通过餐饮文化对比，增强学生的爱国情怀和民族自豪感。

4.专业人士介入，教材内容科学、前沿

本书在参阅国内外烹饪英语教材和深入企业调研的基础上编写而成。编写过程中，行业专家和一线中餐教师在专业知识及单元布局等方面给予了直接指导，保证了教材内容的科学性、前沿性。

建议每个单元安排每周2学时，预计安排18周，36学时。

本书由北京市朝阳区教育科学研究院张艳红和北京市劲松职业高中赵静担任主编，北京市劲松职业高中向军、牛京刚、刘欣雨任副主编，北京市劲松职业高中钱俊、姜珊、夏孟、李蕊、葛苗苗、宋洋、孙敏担任参编。具体编写分工如下：钱俊编写 Unit 1，姜珊编写 Unit 2，赵静编写 Unit 3，刘欣雨编写 Unit 4，李蕊编写 Unit 5，夏孟编写 Unit 6，宋洋编写 Unit 7，葛苗苗编写 Unit 8，张艳红编写 Unit 9，孙敏编写 Unit 10。

本书在修订过程中，对原教材中的题目表述、图片、文字表述以及个别题型等方面进行了修改。同时，根据党的二十大精神，全面贯彻党的教育方针，落实立德树人根本任务，加强推进文化自信，将践行社会主义核心价值观，繁荣发展烹饪服务行业，增强中华传统文化的传播等内容有机融入教材之中。修订后的教材，配备了电子教学资源，学生扫码即可收听单词、词组、对话等，方便学习。

本书在编写过程中，得到了北京市劲松职业高中领导和中餐烹饪协会领导的大力支持，得到了北京市劲松职业高中烹饪专业向军、贾亚东、成晓春和史德杰等老师的大力帮助。本书的出版，也融入了重庆大学出版社编辑的心血，在此一并表示衷心的感谢。

由于编者水平有限，书中难免有不妥之处，还望广大读者批评指正。

编　者
2023年8月

# 前言

（第1版）

随着我国经济的不断发展，尤其是在北京成功举办了2008年奥运会之后，世界各国的游客不断涌入中国，极大地促进了旅游业的发展。随着旅游业的高速发展，对餐饮行业从业者的英语水平提出了更高的要求。为此，笔者及其编写团队成员走进饭店进行调研，与企业高管、专业人士共同开发了这本适合中等职业教育中餐烹饪专业的英语书。本书也可作为美食爱好者的自学用书和参考用书。

本书具有如下特点。

1. 图文并茂，易学易记

本书以学习者为中心，配备了大量彩色图片，图文并茂，浅显直观，易学易记。本书设计的生词、词组、对话内容，能够紧密联系厨房岗位工作实际，达到激发学习者学习兴趣的目的。

2. 构思新颖，实践性强

本书构思新颖，以一名中餐烹饪专业实习生 Jack 到饭店顶岗实习过程中的所见、所学、所感为线索，以情景教学为基石，以活动为内容，强调语言学习的实践性。本书将英语教学与专业教学有机结合，既可作为英语教材又可辅助专业教师实行"双语教学"。

3. 模块设置独特，学习方式灵活自由

每个单元均设置了 Learning Tips 和 My Blog 两个模块。Learning Tips 丰富和补充了相关烹饪文化和厨房操作安全等内容。My Blog 的学习与交流方式，具有鲜明的时代特征，能够提高学生的参与程度和学习兴趣。根据使用者的不同情况，这两个模块的教与学可以采取灵活的方式处理。

4.专业人士介入,教材内容科学、前沿

本书在参阅国内外烹饪英语教材、深入企业调研的基础上编写而成。编写过程中,行业专家和一线中餐教师在专业知识和单元布局等方面给予了直接指导,保证了教材内容的科学性、前沿性。

建议每个单元安排每周2学时,预计安排18周,36学时。

本书由北京市劲松职业高中张艳红、赵静担任主编,重庆市旅游学校杜纲、青岛烹饪职业学校黄艳和北京市劲松职业高中牛京刚担任副主编,北京教育学院英语系教授于淑卿担任主审。主要参编人员及编写章节如下:北京市劲松职业高中钱俊编写 Unit 1,青岛烹饪职业学校黄艳和北京市劲松职业高中贾尚谕编写 Unit 2,北京市劲松职业高中赵静编写 Unit 3,北京市劲松职业高中刘欣雨编写 Unit 4,重庆市旅游学校杜纲编写 Unit 5,北京市劲松职业高中孙文红编写 Unit 6,北京市劲松职业高中王茜编写 Unit 7,北京市劲松职业高中王静编写 Unit 8,北京市劲松职业高中张艳红编写 Unit 9,北京市劲松职业高中郑鸿彦编写 Unit 10。

本书在编写过程中,得到了北京市劲松职业高中学校领导和中餐烹饪协会领导的大力支持,也得到了北京市劲松职业高中烹饪专业向军、贾亚东、成晓春和史德杰等老师的大力帮助。本书的出版,也融入了重庆大学出版社编辑的心血,在此一并表示衷心的感谢。

由于编者水平有限,书中难免有不妥之处,还望广大读者批评指正。

编 者
2015年6月

# Contents

## Unit 1  Kitchen Introduction ........ 1

You will be able to ........ 1
Warming Up ........ 2
Now I Can Learn ........ 3
Now I Can Speak ........ 7
Now I Can Read ........ 8
Now I Can Do ........ 10
Word List ........ 12
Learning Tips ........ 13
Culture Knowledge ........ 13

## Unit 2  Kitchen Equipment ........ 14

You will be able to ........ 14
Warming Up ........ 15
Now I Can Learn ........ 16
Now I Can Speak ........ 20
Now I Can Read ........ 21
Now I Can Do ........ 22
Word List ........ 25
Learning Tips ........ 26
Culture Knowledge ........ 26

## Unit 3  Kitchen Knives ........ 27

You will be able to ........ 27
Warming Up ........ 28
Now I Can Learn ........ 29

# Contents

Now I Can Speak ............................................. 33
Now I Can Read ............................................. 34
Now I Can Do ............................................... 35
Word List ..................................................... 38
Learning Tips ................................................ 39
Culture Knowledge .......................................... 39

## Unit 4  Tools & Utensils .................................. 40

You will be able to ......................................... 40
Warming Up ................................................. 41
Now I Can Learn ............................................ 42
Now I Can Speak ............................................ 46
Now I Can Read ............................................. 47
Now I Can Do ............................................... 49
Word List ..................................................... 51
Learning Tips ................................................ 52
Culture Knowledge .......................................... 52

## Unit 5  Seasonings ....................................... 53

You will be able to ......................................... 53
Warming Up ................................................. 54
Now I Can Learn ............................................ 55
Now I Can Speak ............................................ 59
Now I Can Read ............................................. 60
Now I Can Do ............................................... 62
Word List ..................................................... 64
Learning Tips ................................................ 65
Culture Knowledge .......................................... 65

# Contents

## Unit 6 — Fruits & Nuts — 66

- You will be able to ………… 66
- Warming Up ………… 67
- Now I Can Learn ………… 68
- Now I Can Speak ………… 72
- Now I Can Read ………… 73
- Now I Can Do ………… 75
- Word List ………… 77
- Learning Tips ………… 78
- Culture Knowledge ………… 79

## Unit 7 — Vegetables — 80

- You will be able to ………… 80
- Warming Up ………… 81
- Now I Can Learn ………… 82
- Now I Can Speak ………… 86
- Now I Can Read ………… 87
- Now I Can Do ………… 90
- Word List ………… 92
- Learning Tips ………… 92
- Culture Knowledge ………… 93

## Unit 8 — Meat — 94

- You will be able to ………… 94
- Warming Up ………… 95
- Now I Can Learn ………… 96
- Now I Can Speak ………… 100

# Contents

Now I Can Read ................................................ 101
Now I Can Do .................................................. 104
Word List ..................................................... 106
Learning Tips ................................................. 107
Culture Knowledge ............................................. 107

## Unit 9  Seafood ............................................ 108

You will be able to .......................................... 108
Warming Up ................................................... 109
Now I Can Learn .............................................. 110
Now I Can Speak .............................................. 114
Now I Can Read ............................................... 115
Now I Can Do ................................................. 117
Word List .................................................... 120
Learning Tips ................................................ 121
Culture Knowledge ............................................ 121

## Unit 10  Staple Food & Snacks .............................. 122

You will be able to .......................................... 122
Warming Up ................................................... 123
Now I Can Learn .............................................. 124
Now I Can Speak .............................................. 128
Now I Can Read ............................................... 129
Now I Can Do ................................................. 131
Word List .................................................... 133
Learning Tips ................................................ 134
Culture Knowledge ............................................ 135

参考文献 ..................................................... 136

# Unit 1
# Kitchen Introduction

## You will be able to:

✧ know of some job titles in the kitchen;
✧ describe jobs in the kitchen;
✧ raise the awareness of personal hygiene and kitchen safety.

# Warming Up（热身）

## ❶ Self-introduction（自我介绍）

Hello, I am Li Lei. My English name is Jack. I'm from Beijing Xinhua Vocational School. I'll be working as a trainee in the Grand Hotel for a year. My chef is Mr. Wang. He is very kind and strict. Now, follow me to the kitchen, please.

## ❷ Look and match（看图并连线）

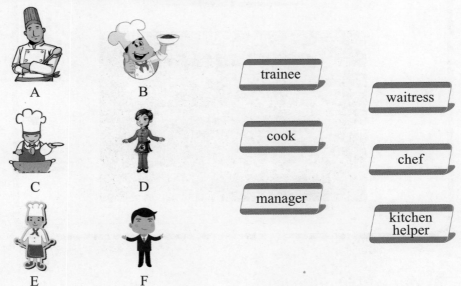

## ❸ Write down the job titles（写出工作岗位的英文表达）

1. _____ The person who cooks food.

2. _____ The person who is the head of cooks.

3. _____ The person who is in charge of the hotel.

4. _____ The person who serves the guests.

# Now I Can Learn（学一学）

 **A**

**1 Read and write**（朗读并写出下列工作岗位的英文表达）

> executive chef 行政总厨师长　　executive sous-chef 行政副总厨师长
> cold kitchen chef 冷菜厨师　　　wok chef 炒锅厨师
> pastry chef 中式面点厨师　　　　chop cook 砧板厨师
> water table cook 水台厨师　　　　assistant cook 打荷厨师
> steamer 上杂工

1._____  2._____  3._____  4._____

5._____  6._____  7._____  8._____

**2 Drill**（句型训练）

**A** What does he do in the kitchen?　　**B** He is a chop cook.

**A** What is he in charge of?　　**B** He is in charge of chopping vegetables and meat.

**3 Listen, complete and read.**（听录音，完成并朗读对话）

**A** what he does in the kitchen　　**B** He is a chop cook

**C** let me show you around　　**D** in charge of chopping vegetables and meat

Unit 1　Kitchen Introduction

**Chef:** Good morning, Jack. Welcome to our hotel.

**Jack:** Good morning. Nice to meet you.

**Chef:** It's your first day here, _____. This way, please. Here is our Executive Chef office, Executive Sous-chef office, Cold kitchen and Hot kitchen. Oh, this is Mr. Smith. _____.

**Jack:** Could you tell me _____?

**Chef:** He is _____. It's a very important job in Chinese kitchen.

**Jack:** I see. Thank you very much.

**Chef:** You are welcome.

## Part B

### 1 Listen and read（听录音，读词组）

wet (one's) hands 湿手　　rub (one's) hands 搓手　　rinse (one's) hands 冲手
hold some water 捧水　　dry (one's) hands 擦干手

### 2 Look, write and talk （看图写英文，并进行对话练习）

A: What should we do before preparing food?
B: We should _____ our hands.

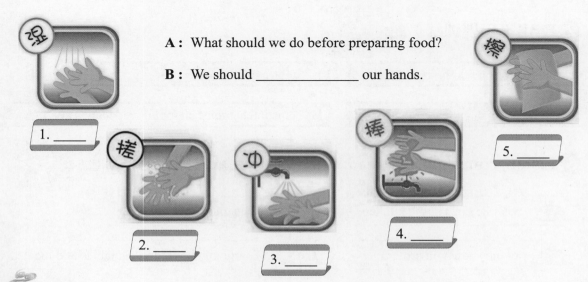

1. _____
2. _____
3. _____
4. _____
5. _____

## 3 Read（读重点词汇）

1. wet hands with soap 湿手抹肥皂
2. rub hands for 20 seconds 搓手 20 秒
3. rinse hands with water 用清水冲净双手
4. hold water to clean and turn off the tap 捧水冲并关水龙头
5. dry hands with a clean towel 用干净毛巾擦干手

Five steps of washing hands

## 4 Choose and read（选择正确答案，并朗读对话）

**Jack:** Excuse me, sir, what should / shall I do before cooking?
**Chef:** You must wash your hands. First you can wet hands with soap.
　　　　Then_____ .
**Jack:** What shall I do next?
**Chef:** After that you can_____, and _____.
**Jack:** What shall I do at last?
**Chef:** Finally you can_____.

A. rinse hands with water
B. dry hands with a clean towel
C. rub hands for 20 seconds
D. hold water to clean and turn off the tap

## Part C

### 1 Read and match（读词组并连线）

clean sinks 清洁水槽　　knife safety 安全用刀　　fire prevention 防止火灾
electric safety 安全用电　　gas prevention 防止漏气

Unit 1　Kitchen Introduction

## 2 Match（连线）

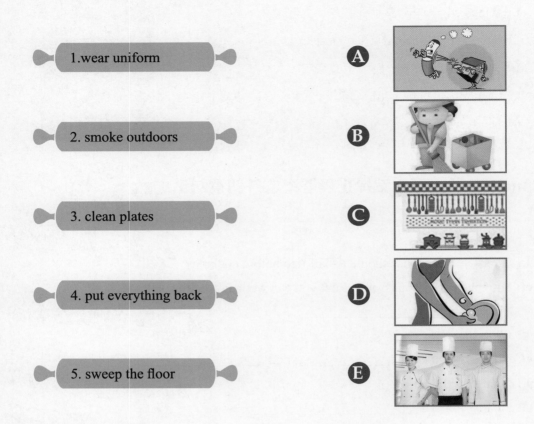

1. wear uniform
2. smoke outdoors
3. clean plates
4. put everything back
5. sweep the floor

A
B
C
D
E

## 3 Fill in and evaluate（补全单词，并进行自我评价）

1. ch_f    2. st_ _ mer    3. sw_ _p    4. tr_ _ nee    5. un_form

6. s_fety    7. g_s    8. sm_ke    9. p_ stry    10. kitch_n

Assessment: If you can write 8–10 words, you are perfect.
If you can write 4–7 words, you are good.
If you can write 1–3 words, you need to try again.

# Now I Can Speak(说一说)

## Make dialogues(对话练习)

**A:** What do you do in the kitchen?
**B:** I am a *wok chef*.
**A:** What do you do as a wok chef?
**B:** I am responsible for *cooking hot dishes*.

waiter / serving the guests
water table cook / preparing fish and seafood
chop cook / chopping

**A:** May I *smoke* in the kitchen?
**B:** No, you mustn't.

put stack box on the floor | take food home | wear shorts | keep the kitchen door open all the time

# Now I Can Read（读一读）

### 1 Read and answer（读对话，回答问题）

**Jack:** Excuse me, when should I wash my hands?
**Chef:** Before preparing food, starting work, and after smoking or using the toilet.
**Jack:** Do I need to wear the cleaned uniform every day?
**Chef:** Sure, you need to do so, and you have to clean the equipment regularly.
**Jack:** Oh, I see. What else should be done?
**Chef:** Dish sinks and surrounding areas should be cleaned.
**Jack:** Yes, I think it's necessary to sweep and mop the floor.
**Chef:** Well done.

1. What should we do before preparing food?
   _____

2. Do I need to wear the cleaned uniform every day?
   _____

3. Do I have to clean the equipment regularly?
   _____

4. What else should be done?
   _____

### 2 Write（参考上一题，用英文写出 Jack 的厨房工作）

1. 洗手　　　　2. 穿干净制服　　3. 清洗厨房设备　　4. 清洗水池　　5. 扫地

_____　　_____　　_____　　_____　　_____

## ❸ Tick（勾出厨师的厨房工作）

☐ play tennis　　☐ clean the floor　　☐ eat food　　☐ wash hands

☐ wipe table　　☐ cook fish　　☐ cut meat　　☐ chat with friends

## ❹ Read and decide（读短文，判读正误）

### Safety in the kitchen

1. All cuts must be bandaged（缠着绷带）with waterproof protectors and gloves should be worn.
2. Kitchen staff with sore throats（嗓子疼）or any diseases（传染病）shall not be permitted to work in the kitchen.
3. No eating or drinking in the kitchen area. No use of tobacco products in the kitchen.

1. _____ All cuts must be bandaged with waterproof protectors.

2. _____ Kitchen staff with sore throats shall be permitted to work.

3. _____ No eating or drinking permitted in the kitchen area.

# Now I Can Do（做一做）

**Exercise 1　Translate into Chinese or English**（单词和词组中英文互译）

1. trainee _____
2. 炒锅厨师 _____
3. chop chef _____
4. 上杂工 _____
5. dish sink _____
6. 洗手 _____
7. pastry chef _____
8. 清洁水槽 _____
9. gas prevention _____
10. 防止火灾 _____

**Exercise 2　Tick off the odd words**（勾出不同类别的单词或词组）

1. A. wet hands　　　B. rub hands　　　C. rinse hands　　　D. hold water
2. A. watch TV　　　B. fire prevention　　C. electric safety　　D. gas prevention
3. A. wok chef　　　B. doctor　　　　　C. trainee　　　　　D. steamer
4. A. pastry chef　　 B. chop chef　　　　C. steamer　　　　 D. wok chef
5. A. glove　　　　　B. uniform　　　　　C. cook　　　　　　D. shorts

**Exercise 3　Make sentences in right order**（连词成句）

1. am responsible, for steaming, I, steamed bread
   _____

2. think, to sweep and mop, is, it, I, necessary, the floor
   _____

3. 20 seconds, hands, for a minimum of, must be washed
   _____

4. waterproof protectors, must be bandaged, all cuts, with
   _____

5. smoke, you, to, in the kitchen, are not allowed
   _____

## Exercise 4  Find the answers（找答案）

1. What hotel do you work in?
2. When should I wash my hands?
3. What do you do in the kitchen?
4. May I smoke in the kitchen?
5. What else should be done?

A. After using the toilet.
B. I work in the Great Wall Hotel.
C. Dish sink and surrounding areas should be cleaned.
D. No, you mustn't.
E. I am a pastry sous-chef.

## Exercise 5  Write（写出你/你们未来的职业规划）

 **e.g.** What would you like to be in the future?
I'd like to be *a hot kitchen sous-chef.*

• I want to be a /an _____ .

• We want to be _____ .

## Exercise 6  Translate into Chinese or English（句子中英文互译）

1. 我在长城饭店工作。
2. 炒锅厨师负责做热菜。
3. 你在厨房做什么工作?
4. Dish sinks and surrounding areas should be cleaned.
5. Nobody is allowed to smoke in the kitchen.

# Word List（单词表）

| | | | | | |
|---|---|---|---|---|---|
| trainee | [ˌtreiˈniː] | 实习生 | **hold** water | [həuld] | 捧水 |
| serve | [səːv] | 服务 | **dry** hands | [drai] | 擦干手 |
| **executive** chef | [igˈzekjutiv] | 行政总厨师长 | soap | [səup] | 肥皂 |
| cold **kitchen** chef | [ˈkitʃin] | 冷菜厨师 | clean **sink** | [siŋk] | 清洁水槽 |
| **wok** chef | [wɔk] | 炒锅厨师 | knife **safety** | [ˈseifti] | 安全用刀 |
| executive **sous-chef** | [ˈsuːʃef] | 行政副总厨师长 | fire **prevention** | [priˈvenʃn] | 防止火灾 |
| **pastry** chef | [ˈpeistri] | 中式面点厨师 | electric safety | [iˈlektrik] | 安全用电 |
| **chop** cook | [tʃɔp] | 砧板厨师 | **gas** prevention | [gæs] | 防止漏气 |
| steamer | [ˈstiːmə(r)] | 上杂工 | wear **uniform** | [ˈjuːnifɔːm] | 穿制服 |
| **water** table cook | [wɔːtə] | 水台厨师 | **keep**…open | [kiːp] | 开着 |
| **assistant** cook | [əˈsistənt] | 打荷厨师 | equipment | [iˈkwipmənt] | 设备 |
| guest | [gest] | 客人 | regularly | [ˈregjələli] | 有规律地 |
| be in **charge** of | [tʃɑːdʒ] | 负责 | surrounding | [səˈraundiŋ] | 周围 |
| be **responsible** for | [riˈspɔnsəbl] | 为……负责 | areas | [ˈeəriəz] | 地区 |
| **wet** hands | [wet] | 湿手 | staff | [stɑːf] | 员工 |
| **rub** hands | [rʌb] | 搓手 | wipe | [waip] | 擦去 |
| **rinse** hands | [rins] | 冲手 | permit | [pəˈmit] | 允许 |

## Learning Tips（学习提示）

我们学习了厨房里的各种职位名称，你知道酒店对食品安全管理的规定吗？食品安全是指食品无毒、无害，符合应当有的营养要求，对人体健康不造成任何急性、亚急性或者慢性危害。另外，酒店从业人员必须持有国家颁发的"健康证"。

 ## Culture Knowledge（文化知识）

### 好厨师的要求

**厨德是当好厨师的根本**

优秀的厨师必须具备良好的厨德，培养良好的厨德才能使厨师走向成功，在行业内有所建树。

**拥有精湛的技艺**

厨艺是厨师立足的关键。要成为一名优秀的厨师，必须拥有精湛的技艺。比如一名优秀的湘菜厨师，不仅要精通湘菜，还必须旁通其他菜系。

**对企业和顾客充满感情**

这里谈到的感情就是在工作中要把促进发展和满足顾客需求摆在首位，将个人利益置后。带着感情工作，这是一名合格的厨师不可或缺的品质。

# Unit 2
# Kitchen Equipment

## *You will be able to:*

- ✧ get familiar with the equipment in the kitchen;
- ✧ say the names of equipment in the kitchen;
- ✧ describe the function of auxiliary equipment.

# Warming Up (热身)

**1** Look and match (看图并连线)

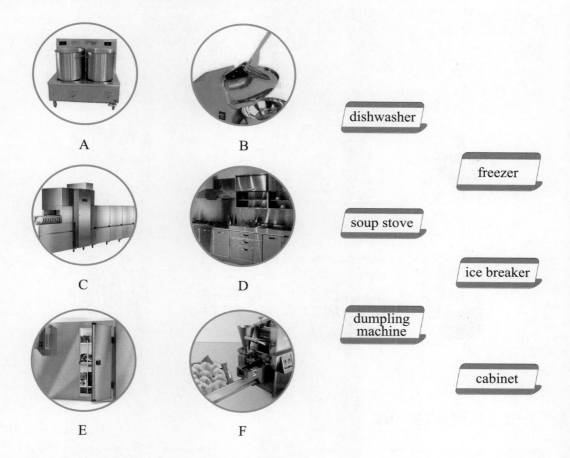

**2** Decide true or false (判断正误)

(　　) 1. An ice breaker is used for breaking ice.

(　　) 2. A freezer is often used for keeping tableware.

(　　) 3. A dishwasher is mostly used to wash vegetables.

(　　) 4. A soup stove is a kind of heating cookware in the kitchen.

(　　) 5. A dumpling machine is used for making dumplings.

# Now I Can Learn (学一学)

## Part A

**1** Read and write (读词组，写出下列设备的英文表达)

> steam cabinet 蒸柜　　　　　　ice machine 制冰机
> refrigerated cabinet 冷藏柜　　 boiling plate 电热器
> gas stove 煤气灶　　　　　　　oven 烤箱
> steam jacket cooker 蒸汽夹层锅　soup stove 汤炉

1._____　2._____　3._____　4._____

5._____　6._____　7._____　8._____

**2** Drill (句型训练)

**A** What do we call this machine in English?　**B** It's an ice machine.

**A** What is it used for?　**B** It is used for making ice.

**3** Listen, complete and read (听录音，完成并朗读对话)

 What is it used for　　　 What is that called in English

 It's very useful　　　　　**D** What is this machine

16　中餐烹饪英语

**Chef:** Let's get to know some equipment today.
**Jack:** OK. _____, chef?
**Chef:** It's a refrigerated cabinet.
**Jack:** _____?
**Chef:** It is used for keeping food fresh.
**Jack:** _____?
**Chef:** It is called an ice machine.
**Jack:** Is it used for making ice?
**Chef:** Yes, especially in summer.
**Jack:** _____.
**Chef:** I agree with you.

## Part B

### 1 Listen and read（听录音，读词组）

flapjack machine 烙饼机     liquidizer 榨汁机         noodle press 压面机
bread slicer 面包切片机    fermentation machine 饧发机    slicer 切片机
meat grinder 绞肉机        blender 搅拌机；料理机         mixer 搅拌机

### 2 Look, write and talk（看图写英文，并进行对话练习）

A: What do we call this machine in English?
B: It's called a _____.

1. _____

5. _____

2. _____   3. _____   4. _____   6. _____

7. _____   8. _____   9. _____

## 3 Read（读重点词汇）

1. grind meat 绞肉
2. liquidize fruits and vegetables 榨果蔬汁
3. slice bread / meat 将面包 / 肉切片

## 4 Choose and read（选择正确答案，并朗读对话）

A. slice it very well

Jack: I need to grind some meat, chef. What kind of machine should I use?
Chef: _____ .
Jack: What shall I do with these oranges?
Chef: Put them in the liquidizer. It can_____.
Jack: _____?
Chef: Use the bread slicer. It can_____.
Jack: OK, thank you, chef.

B. A meat grinder

C. liquidize fruits and vegetables

D. How about the bread

# Part C

## 1 Read and match（读词组并连线）

cooker hood 排油烟机       motor-operated fan 电动排风扇

disinfection cabinet 餐具消毒柜       container cleaning machine 容器清洗机

high pressure spray washing machine 高压喷射洗涤机

   A            B            C            D            E

## 2 Match（连线）

1. liquidize oranges    A

2. grind meat    B

3. slice meat    C

4. make ice    D

5. keep vegetables    E

## 3 Fill in and evaluate（补全单词，并自我评价）

1. sl_cer    2. l_qu_d_zer    3. fr_ _zer    4. bl_nd_r    5. ice br_ _ker

6. g_s st_ve    7. d_shw_sher    8. cab_n_t    9. mach_ne    10. m_ _t gr_nder

Assessment: If you can write 8–10 words, you are perfect.
             If you can write 4–7 words, you are good.
             If you can write 1–3 words, you need to try again.

# Now I Can Speak（说一说）

## Make dialogues（对话练习）

e.g.
**A:** What shall I do with the *meat*?
**B:** *Grind it (them).*
**A:** With what?
**B:** *With a meat grinder.*

bread / slice / bread slicer
mangoes / liquidize / liquidizer
dishes / wash / dishwasher
potatoes / fry / gas stove

e.g.
**A:** What do you call this, chef?
**B:** It is called a(n)_____.
**A:** What is it used for?
**B:** It is used for _____.
**A:** Can you show me how to use it?
**B:** Yes, of course.

ice machine / making ice    dumpling machine / making dumplings    noodle press / pressing noodles    meat grinder / grinding meat

# Now I Can Read（读一读）

## ❶ Read and answer（读对话，回答问题）

**Chef :** Today, I'll show you around our kitchen.
**Jack :** Really? That's great.
**Chef :** OK, let's look at the stove. It's called a gas stove.
**Jack :** Wow! It's so tidy. How about that?
**Chef :** It's an advanced cooker hood.
**Jack :** It can exhaust oil smoke, right?
**Chef :** Yes, you are right.
**Jack :** Is it a freezer?
**Chef :** Yes, we usually use it to keep meat and seafood fresh.
**Jack :** I see. What's this machine?
**Chef :** It's a dishwasher. It can help us to save time.
**Jack :** That's great! And what's that?
**Chef :** That is an ice machine.
**Jack :** Oh, I learn a lot today. Thank you.

1. What is a freezer used for?
   _____

2. What is a cooker hood used for?
   _____

3. Which machine can wash dishes?
   _____

4. Can you list some pieces of equipment? (At least 4.)
   _____

## ❷ Write（根据描述写出设备的英文名称）

1. _____ It is used for cooking dishes.

2. _____ It can keep meat and seafood fresh.

3. _____ It can exhaust oil smoke.

4. _____ It can make ice.

5. _____ It can wash plates and bowls.

### ❸ Tick（勾出对话中提到的厨房设备）

☐ dishwasher  ☐ ice machine  ☐ liquidizer

☐ freezer  ☐ cooker hood  ☐ gas stove

### ❹ Match（连线）

| soup stove | clear soup |
| meat grinder | oranges |
| liquidizer | beef |
| noodle press | ice |
| steam cabinet | noodles |
| ice breaker | mantou |

## Now I Can Do（练一练）

**Exercise 1**  Translate into Chinese or English（单词和词组中英文互译）

1. slicer _____    2. 洗碗机 _____

3. soup stove _____    4. 压面机 _____

5. ice machine _____    6. 煤气灶 _____

7. steam cabinet _____   8. 冷藏柜_____

9. cooker hood _____   10. 电热器_____

## Exercise 2   Choose the best answer（选择正确答案）

1. I like juice, so I use the _____ to make juice every day.
   A. slicer          B. liquidizer        C. grinder         D. freezer

2. This is a _____. It can get rid of the oil smoke from the kitchen.
   A. steam cabinet   B. dishwasher        C. freezer         D. cooker hood

3. _____ is a machine to make noodles.
   A. Slicer          B. Blender           C. Noodle press    D. Freezer

4. In order to keep the food fresh, we'd better put them into a _____.
   A. blender         B. dishwasher        C. grinder         D. freezer

5. I need to use a _____ to prepare some meat for dumplings.
   A. meat grinder    B. liquidizer        C. soup stove      D. freezer

## Exercise 3   Look and write（根据加工图片写出所需厨房设备的英文表达）

## Exercise 4   Make sentences in right order（连词成句）

1. machine, we, do, call, what, this

_____

Unit 2   Kitchen Equipment

2. with, what, carrots, do, I, shall, these

_____

3. oranges, it, use, to, can, liquidize, you

_____

4. you, how, grind, me, can, show, meat, to

_____

5. can, dishwasher, these, I, dishes, use, to, wash

_____

## Exercise 5    Complete the sentences（完成句子）

| gas stove    freezer    dishwasher    dumpling machine    soup stove |

1. We can use _____ to cook food.
2. _____ can wash plates and bowls.
3. We can use _____ to make soup.
4. If we want to keep fruits and vegetables fresh, we must put them into _____ .
5. _____ can help us make dumplings.

## Exercise 6    Translate into Chinese or English（句子中英文互译）

1. 压面机可以用来制作面条。
2. 汤姆正在用榨汁机榨果汁。
3. What do we call this machine in English?
4. Can you show me how to use it?
5. The freezer is used to keep meat and seafood fresh.

# Word List（单词表）

| | | | | | |
|---|---|---|---|---|---|
| refrigerated cabinet | [ri'fridʒəreitid 'kæbinət] | 冷藏柜 | grind | [graind] | 粉碎/研磨 |
| ice **breaker** | ['breikə(r)] | 碎冰器 | blender | ['blendə(r)] | 搅拌机；料理机 |
| ice machine | ['ais mə'ʃi:n] | 制冰机 | blend | [blend] | 搅拌 |
| boiling plate | ['bɔiliŋ pleit] | 电热器 | mixer | ['miksə(r)] | 搅拌机 |
| soup stove | [su:p stəuv] | 汤炉 | liquidizer | ['likwidaizə(r)] | 榨汁机 |
| oven | ['ʌvn] | 烤箱 | liquidize | ['likwidaiz] | 把水果和蔬菜等榨成汁 |
| **steam jacket** cooker | [sti:m 'dʒækit] | 蒸汽夹层锅 | cooker hood | ['kukər hud] | 排油烟机 |
| dishwasher | ['diʃˌwɔʃə(r)] | 洗碗机 | motor-operated | ['məutə(r) 'ɔpəreitid] | 电动的 |
| **dumpling** machine | ['dʌmpliŋ] | 饺子机 | **disinfection** cabinet | [ˌdisin'fekʃən] | 餐具消毒柜 |
| **flapjack** machine | ['flæpdʒæk] | 烙饼机 | high **pressure spray** washing machine | ['preʃə sprei] | 高压喷射洗涤机 |
| noodle press | ['nu:dl pres] | 压面机 | freezing compartment | ['fri:ziŋ kəm'pɑ:tmənt] | 冷冻室 |
| slicer | ['slaisə(r)] | 切片机 | **fermentation** machine | [ˌfə:men'teiʃən] | 饧发机 |
| steam cabinet | [sti:m 'kæbinət] | 蒸柜 | **container** cleaning machine | [kən'teinə(r)] | 容器清洗机 |
| grinder | ['graində(r)] | 粉碎/研磨机 | | | |

# Learning Tips（学习提示）

我们学习了厨房里的各种设备，你知道酒店对厨房设施设备的管理有哪些规定吗？希望你从现在开始注意了解，以便今后能出色地胜任工作。加油！

## 酒店厨房设施设备管理制度

1. 厨房设备，如绞肉机、冰箱、蒸柜、压面机等设备均由专人使用。
2. 不经过厨师长的同意，不得擅自使用厨房设备。
3. 定期对自己使用的设备进行维护、保养，确保设备的正常使用。
4. 下班后，厨师长要安排专人对厨房所有设备及电源进行检查，确保万无一失后方可离开厨房，并锁好厨房门锁。
5. 发现故障隐患，要及时向厨师长汇报，及时报修，由专业人员进行维修。

# Culture Knowledge（文化知识）

## 厨房设备的分类

按照用途来分，厨房设备可分为以下5类。

第一类是储物设备，分为食品储藏和器物用品储藏两大部分。食品储藏又分为冷藏和非冷藏。冷藏是通过厨房内的电冰箱、冷藏柜等实现的。器物用品储藏是为餐具、炊具、器皿等提供存储空间的。储藏功能是通过各种底柜、吊柜、角柜、多功能装饰柜等完成的。

第二类是洗涤设备，包括冷热水的供应系统、排水设备、洗物盆、洗物柜等。洗涤后在厨房操作中产生的垃圾，应设置垃圾箱或卫生桶等，酒店厨房还应配备消毒柜。

第三类是调理设备，主要包括调理的台面，整理、切菜、配料、调制的工具和器皿，如厨房用切片机、榨汁机、绞肉机等。

第四类是烹调设备，主要有炉具、灶具和烹调时的相关工具和器皿。现在，电饭锅、高频电磁灶、微波炉、烤箱等也大量进入厨房。

第五类是进餐设备，主要包括进餐时的设备和器皿等。

# Unit 3
# Kitchen Knives

## You will be able to:

- say the names of different kinds of knives;
- get familiar with the usages of different kinds of knives;
- make short conversations on talking about knives.

# Warming Up（热身）

## 1 Look and match（看图并连线）

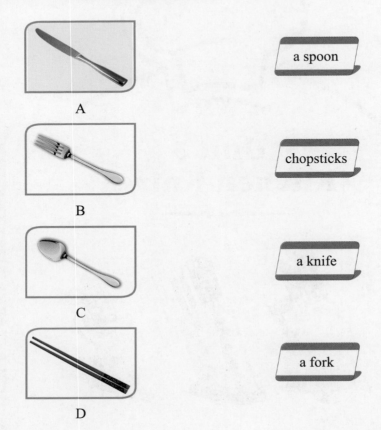

## 2 Decide true or false（判断正误）

1._____ We usually use a knife to mix ingredients.

2._____ Chinese are used to pick up food with chopsticks.

3._____ We should clean the knife immediately after using it.

4._____ The bread knife is a Chinese knife.

5._____ We should use different knives to cut different things.

# Now I Can Learn（学一学）

**Part A**

### 1 Read and write（读单词和词组，写出下列刀具的英文表达）

> chopping knife 文武刀　　slicing knife 片刀
> carving knife 雕刻刀　　cleaver 砍刀
> straight cutting 直切　　pull cutting 拉切
> saw cutting 锯切

1._____　　2._____　　3._____　　4._____

### 2 Drill（句型训练）

**A** Which knife shall I use?

**B** You can use a chopping knife.
You should use your chopping knife.

**A** Do you know the different ways of cutting?

**B** Yes. I've learned straight cutting, pull cutting and saw cutting in my school.

### 3 Listen, complete and read（听录音，完成并朗读对话）

**A** Chinese style knives are heavier

**B** cleaver and slicing knife

**C** the different ways of cutting

**D** Chinese style knives and Western style knives

**Jack:** Wow, there are so many knives. What are they?
**Chef:** They are _____.
**Jack:** What are the differences between them?
**Chef:** Well, _____.
**Jack:** What kind of knives do we often use?
**Chef:** We usually use chopping knife, _____. Do you know _____?
**Jack:** Yes, I've learned straight cutting, pull cutting and saw cutting in my school.
**Chef:** That's wonderful.

*Notes:* 刀工中的切法有很多种，除直切、拉切和锯切外，还有推切、推拉切、侧切和滚切等。

## Part B

### 1 Listen and read（听录音，读单词和词组）

| | | |
|---|---|---|
| lamb knife 羊肉刀 | kitchen scissors 厨房剪 | peeler 削皮刀 |
| scaler 刮鳞器 | roast duck knife 烧鸭刀 | melon baller 挖球器 |
| whetstone 磨刀石 | steel 磨刀棒 | |

### 2 Look, write and talk（看图写英文，并进行对话练习）

A：What's in your hand?
B：It's a _____ .

1. _____
2. _____
3. _____
4. _____
5. _____
6. _____
7. _____
8. _____

## ③ Read（读重点词汇）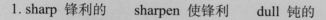

1. sharp 锋利的    sharpen 使锋利    dull 钝的
2. cut up... with 用……切
3. special job 特殊用途，特殊工作
4. go ahead 可以，行

## ④ Choose and read（选择正确答案，并朗读对话）

Jack: Shall I peel the potato with _____ ?
Chef: Go ahead.
Jack: Wow, this chopping knife is very sharp. Is it _____ ?
Chef: No. It is for many different things.
Jack: Can I cut _____ with my chopping knife?
Chef: Well, you'd better use your roast duck knife. It is _____ .
Jack: En, it cuts very well. I really enjoy it.

A. roast duck
B. this peeler
C. easier to use
D. for special jobs

## Part C

### ① Read and tick（读词组，勾出你用过的刀具）

chef's knife 厨刀    cheese knife 奶酪刀    paring knife 水果刀
salmon knife 三文鱼刀    boning knife 剔骨刀    oyster knife 牡蛎刀

Unit 3  Kitchen Knives  31

## 2 Match（连线）

1. chop the meat

 A

2. dice the onion

 B

3. shred the potato

 C

4. slice the salmon

 D

5. cut the vegetable

 E

## 3 Fill in and evaluate（补全单词，并自我评价）

1. ch_p    2. sl_ce    3. cl_ _ver    4. sh_ _p    5. p_ _ler

6. d_ce    7. b_ne    8. _ _eese    9. r_ _ st    10. a_ ead

Assessment: If you can write 8–10 words, you are perfect.
　　　　　　If you can write 4–7 words, you are good.
　　　　　　If you can write 1–3 words, you need to try again.

# Now I Can Speak (说一说)

## Make dialogues (对话练习)

**e.g.**
**A:** What's this?
**B:** This is a *slicing knife*.
 I am *cutting the vegetables* with it.

cleaver / chop / meat
oyster knife / open / oyster
cheese knife / cut / cheese

**e.g.**
**A:** What are we going to do?
**B:** We are going to *cut some oranges* with *paring knife*.

| cut potatoes / slicing knife | scale fish / scaler | slice salmon / salmon knife | chop chicken / chopping knife |

# Now I Can Read (读一读)

## ❶ Read and answer (读对话，回答问题)

> Jack: My knife is dull. It doesn't work well.
> Chef: Use a steel.
> Jack: Where can I find it?
> Chef: You can use mine. Here you are.
> Jack: Thank you. What else can I use to sharpen my knife?
> Chef: You may use a whetstone, or you can also use an electric knife sharpener. It is safer.
> Jack: Oh, I see. Thank you for telling me.

1. What's wrong with Jack's knife?
   _____

2. What can Jack use to sharpen his knife?
   _____

3. Which kind of knife sharpener is safer?
   _____

## ❷ Write (参考上一题，用英文写出正确的磨刀方式)

1. _____　　2. _____　　3. _____

## 3 Read and Tick (读用刀安全须知，勾选出正确图片)

Keep the knife on a stable (平稳的) cutting board.
Store the knives securely (安全地) after use.
Pass a knife with the knife pointing downwards (朝下).
Do not play with a knife.
Do not stick a knife on the cutting board.
Do not touch the knife edge (刀刃) with your finger.

# Now I Can Do（练一练）

Exercise 1　Translate into Chinese or English（单词和词组中英文互译）

1. sharp _____　　2. 砍刀 _____

3. dull _____　　4. 片刀 _____

5. steel _____		6. 挖球器 _____
7. scaler _____		8. 羊肉刀 _____
9. fork _____		10. 水果刀 _____

## Exercise 2   Tick off the odd words（勾出不同类别的单词或词组）

1. A. chopsticks    B. scissors         C. spoon           D. fork
2. A. meat          B. apple            C. salmon          D. chop
3. A. cut           B. chop             C. slice           D. scale
4. A. sharp         B. dull             C. onion           D. safe
5. A. lamb knife    B. cheese knife     C. oyster knife    D. salmon knife

## Exercise 3   Find the answers（找答案）

1. What knife shall I use?
2. What's in your hand, Mr. Smith?
3. Shall I peel the potato with this peeler?
4. What are we going to do next?
5. Where can I find a steel?

A. The steel is on the shelf.
B. Go ahead.
C. You can use the chopping knife.
D. It's a carving knife.
E. We are going to cut some apples with the paring knife.

## Exercise 4   Make sentences in right order（连词成句）

1. things, should, different, use, knives, to, we, cut, different

   _____

2. them, what, are, the, between, differences

   _____

3. is, your, in, what, hand

_____

4. am, I, cutting, slicing knife, the, vegetables, a, with

_____

5. can, knife, I, use, what, to, sharpen, my

_____

## Exercise 5  Classify and write（分类并写出下列刀具）

salmon knife    chopping knife    lamb knife    oyster knife
slicing knife   cheese knife      paring knife  cleaver

## Exercise 6  Translate into Chinese or English（句子中英文互译）

1. 文武刀的用途很多。

2. 我正打算切洋葱和土豆。

3. 我的砍刀钝了，我该磨刀了。

4. You can chop the bone with your cleaver.

5. Chinese style knife is heavier than Western style knife.

# Word List（单词表）

| | | | | | |
|---|---|---|---|---|---|
| spoon | [spu:n] | 匙，勺子 | steel | [sti:l] | 磨刀棒 |
| chopsticks | ['tʃɔpstiks] | 筷子 | **chef**'s knife | [ʃef] | 厨刀 |
| knife | [naif] | 刀具，餐刀 | **cheese** knife | [tʃi:z] | 奶酪刀 |
| fork | [fɔ:k] | 餐叉，叉子 | **paring** knife | ['peəriŋ] | 水果刀 |
| **chopping** knife | ['tʃɔpiŋ] | 文武刀 | **salmon** knife | ['sæmən] | 三文鱼刀 |
| **slicing** knife | ['slaisiŋ] | 片刀 | **boning** knife | ['bəuniŋ] | 剔骨刀 |
| **carving** knife | ['kɑ:viŋ] | 雕刻刀 | **oyster** knife | ['ɔistə(r)] | 牡蛎刀 |
| cleaver | ['kli:və(r)] | 砍刀 | sharp | [ʃɑ:p] | 锋利的 |
| **straight** cutting | ['streit] | 直切 | sharpen | ['ʃɑ:pən] | 使锋利 |
| **pull** cutting | [pul] | 拉切 | dull | [dʌl] | 钝的 |
| **saw** cutting | [sɔ:] | 锯切 | dice | [dais] | 切块 |
| **lamb** knife | [læm] | 羊肉刀 | shred | [ʃred] | 切丝 |
| kitchen **scissors** | ['sizəz] | 厨房剪 | electric | [i'lektrik] | 电动的 |
| scaler | ['skeilə(r)] | 刮鳞器 | stable | ['steibl] | 平稳的 |
| **roast duck** knife | [rəust dʌk] | 烧鸭刀 | securely | [si'kjuəli] | 安全地 |
| melon baller | ['melən 'bɔ:lə] | 挖球器 | downwards | ['daunwədz] | 朝下 |
| whetstone | ['wetstəun] | 磨刀石 | knife **edge** | [edʒ] | 刀刃 |

# Learning Tips（学习提示）

厨房刀具和我们的生活息息相关，有厨房就离不开厨房刀具。为了安全和身体健康，同时使刀具保持良好的使用性能，应按刀具的使用功能正确使用。一般可以遵循以下方法。

1. 使用砍刀或剔骨刀斩骨时，要直上直下，勿左右摇摆。
2. 刀具使用完毕，用清水清洗，先将刀体和刀柄抹干净，然后放进刀座内。
3. 刀具应放于通风透气、没有酸碱腐蚀的地方。
4. 定期复磨刀具，清除刀具上的污物、油渍、污渍等。
5. 千万不要把刀放在灶具上，否则刀具容易损坏或造成烫伤。

# Culture Knowledge（文化知识）

### 扬州厨刀

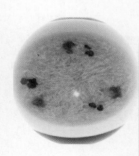

扬州，中国历史文化名城，至今已有2500多年的历史。扬州素来是人文荟萃之地，风物繁华之城。说到扬州，不得不提到天下闻名的"扬州三把刀"：厨刀、修脚刀、理发刀。"三把刀"在扬州人手里不仅是一门技术，更是一门艺术，独具地方特色，是扬州文化的一部分。而最能展现厨刀绝妙刀工的菜肴那就得数文思豆腐羹了。文思豆腐羹是扬州地区的传统名菜，属于淮扬菜。

# Unit 4
# Tools & Utensils

## You will be able to:

◇ get familiar with the cooking utensils in the kitchen;

◇ talk about the usage of cooking utensils;

◇ describe the usage of cooking utensils.

# Warming Up（热身）

**1** Look and match（看图并连线）

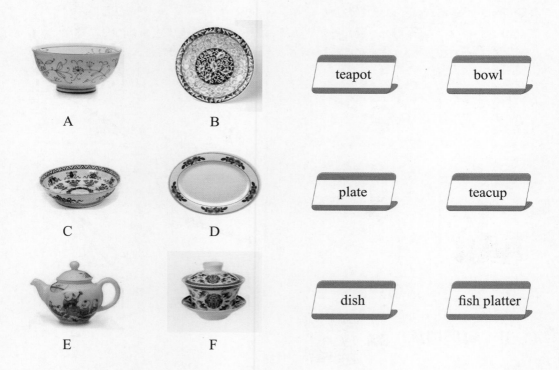

**2** Decide true or false（判断正误）

1._____ Chefs usually put fruit on the fish platter for decoration (装饰).

2._____ The teapot is used for keeping tea hot.

3._____ Dinner starts with a small dish, which is often called a starter.

4._____ Plates are used to hold Chinese stir-fry dishes.

5._____ Bowls are used for holding soup not rice.

# Now I Can Learn（学一学）

## Part A

**1** Read and write（读单词和词组，写出下列工具的英文表达）

| skimmer 漏勺 | rolling pin 擀面杖 | spatula 铲子 | mallet 肉锤 |
| ladle 炒勺 | seasoning pot 调味罐 | soup spoon 汤勺 | chopsticks 筷子 |

1._____  2._____  3._____  4._____

5._____  6._____  7._____  8._____

**2** Drill（句型训练）

**A** What do we call these tools in English?

**B** They are called mallet, skimmer, soup spoon and spatula.

**A** What shall we do with the tools?

**B** Soften the meat with a mallet. Make dumpling wrappers with a rolling pin. Pick up the food with chopsticks.

**3** Listen, complete and read（听录音，完成并朗读对话）

**A** would you fetch some tools for me

**B** we also use it to remove the oil on the surface

**C** we dip up soup from the pot with it

**D** we make dumpling wrappers with them

**Jack:** What would you like me to do now, chef?

**Chef:** _____?

**Jack:** Sure. A skimmer, whisk, spatula and mallet. What can we do with the mallet?

**Chef:** A mallet is used to soften the meat.

**Jack:** OK, I see. What about the soup spoon?

**Chef:** _____, and sometimes _____.

**Jack:** These must be rolling pins. _____.

**Chef:** If you like, you may have a try.

## Part B

### 1 Listen and read (听录音，读单词和词组)

| | | | |
|---|---|---|---|
| pressure cooker 压力锅 | electric cooker 电饭锅 | frying pan 煎锅 | braising pan 焖锅 |
| wok 中式炒锅 | hot pot 火锅 | steamer 蒸锅 | marmite 砂锅 |

### 2 Look, write and talk (看图写英文，并进行对话练习)

1. _____

2. _____

3. _____

4. _____

A: Do you know this tool?

B: Yes, it is called _____.

8. _____

5. _____

6. _____

7. _____

Unit 4 Tools & Utensils

## 3 Read（读重点词汇）

1. over low heat 慢火的
2. steam pastry 蒸糕点
3. fried eggs 煎蛋

## 4 Choose and read（选择正确答案，并朗读对话）

Jack: What kinds of tools are used for cooking hot dishes?
Chef: We usually _____ .
Jack: How about marmites and braising pans?
Chef: They are suitable for _____ .
Jack: What is the steamer for?
Chef: _____ .
Jack: OK, I see. It must be a frying pan, right?
Chef: Yes, you can _____ .

A. low-heated dishes

B. use wok to cook hot dishes

C. make fried eggs with it

D. It is used for steaming pastries

## Part C

### 1 Read and tick（读单词和词组，勾出你常用的工具）

anti-hot folder 防烫夹    colander 沥水篮    grater 擦子
cutting board 切菜板    sieve 筛子

## 2 Match（连线）

1. drain the vegetables — A
2. grate the carrot — B
3. brush the oil — C
4. slice the meat — D
5. pick up the food — E

## 3 Fill in and evaluate（补全单词，并自我评价）

1. gr__ter      2. br__sh      3. s__eve      4. col____der      5. dr__in

6. m__llet      7. p__ck       8. sl__ce      9. c__rrot         10. ch__p

Assessment: If you can write 8–10 words, you are perfect.
If you can write 4–7 words, you are good.
If you can write 1–3 words, you need to try again.

Unit 4  Tools & Utensils

# Now I Can Speak（说一说）

## Make dialogues（对话练习）

e.g.
**A:** How shall I cook the *red beans*?
**B:** *Boil them*.
**A:** With what?
**B:** With *a pressure cooker*.

pastry / Steam it / a steamer
rice / Boil it / an electric cooker
meat / Dip it in boiling water / a hot pot

e.g.
**A:** What can you cook with a frying pan?
**B:** I can cook _____ .

| tofu omelet | fried egg | fried dumplings | fried fish |
| 锅贴豆腐 | 煎蛋 | 煎饺 | 煎鱼 |

## *Now I Can Read*（读一读）

**1** Read and answer（读对话，回答问题）

**Jack:** Chinese wok is one of the most important pieces of cookware in the kitchen. Could you tell me why woks are used so much in Chinese cooking?

**Chef:** Well, Chinese cooks use wok for some reasons. For example, a wok has a large surface area to increase the speed.

**Jack:** Are there any other reasons?

**Chef:** It can keep most vegetables' fresh taste.

**Jack:** Is it so important?

**Chef:** Yes. At the same time, wok's high sides help to keep the food in the rounded bottom, so it can save oil.

**Jack:** I see. Excuse me, sir. Can we use a wok for boiling?

**Chef:** A wok is most commonly used for stir-frying. We steam dumplings or other food with steamers. Soup pots are suitable for stewing and making soup.

**Jack:** I've really learned a lot. Thank you so much.

1. What is one of the most important pieces of cookware in the kitchen?
   _____

2. How many pieces of cookware are mentioned in the dialogue?
   _____

3. What are they?
   _____

4. Can you list any other heating tools you often use in the kitchen? What are they?
   _____

## 2 Write（参考上一题，用英文写出炊具名称）

1._____   2._____   3._____

## 3 Match（将菜肴与所用炊具及烹饪方法进行关联）

1.

A

stir-frying

2.

B

steaming

3.

C

stewing

## 4 Decide true or false（判读正误）

1._____ A Chinese wok has a large surface area to allow a lot of food.

2._____ Chinese cooks never use other pots except woks.

3._____ Soup pots are suitable for stewing and making soup.

4._____ A wok uses less oil than a frying pan and keeps dish fresh.

5._____ We must pay more attention to kitchen safety when we work in it.

# Now I Can Do (做一做)

### Exercise 1   Fill in the blanks (选择填空)

1. Sift the flour with a _____.
2. Drain the carrots in the _____.
3. Beat the meat with a _____.
4. Fry the egg in the _____.
5. Stew the bone in the _____.

- mallet
- marmite
- sieve
- frying pan
- colander

6. _____ the pan off the fire.
7. _____ the meat in the hot pot.
8. _____ the plate with the anti-hot folder.
9. _____ the surface of the oil with a soup ladle.
10. _____ the potato with the grater.

- Boil
- Take
- Grate
- Lift
- Clean

### Exercise 2   Find the answers (找答案)

1. Could you tell me which one is the cream soup cup?
2. If I want a cup of coffee, which one shall I choose?
3. Which one is used for drinking wine?
4. Which utensil is used for holding chopsticks?
5. We drink juice with it.

A. wine glass    B. juice glass    C. coffee cup    D. chopstick holder    E. jar

Unit 4   Tools & Utensils   49

## Exercise 3  Make sentences in right order（连词成句）

1. cut, and, on, the, we, meat, other things, cutting board

   _____

2. steam, the pastries, we, in, the, usually, steamer

   _____

3. a sieve, used , flour, is, for, sifting

   _____

4. we, make, in, soup, the marmite, usually

   _____

5. you, clean, would, help, me, the, dishes, and, plates, to

   _____

## Exercise 4  Translate into Chinese（翻译成中文）

1. cloth（布）
   table cloth(    )    dust cloth(    )    dish cloth(    )
2. pot（罐，壶）
   stew pot (    )    coffee pot (    )    soup pot (    )
3. boiler（汽锅）
   double boiler (    )    steam boiler (    )    rice boiler (    )
4. plate（浅盘，碟）
   square plate(    )    salad plate(    )    dessert plate(    )

## Exercise 5  Translate into Chinese or English（句子中英文互译）

1. 我能用盅来盛汤吗？
2. 请用沥水篮将蔬菜沥干。
3. 我正在用煎锅煎鱼。
4. Clean the surface of the oil with a soup ladle.
5. What shall I do with the steamer?

# Word List（单词表）

| teapot | ['tiːˌpɔt] | 茶壶 | wok | [wɔk] | 中式炒锅 |
|---|---|---|---|---|---|
| bowl | [bəul] | 碗 | steamer | [stimə(r)] | 蒸锅 |
| plate | [pleit] | 盘子 | marmite | ['mɑ(r)mait] | 砂锅 |
| teacup | ['tiːˌkʌp] | 茶杯 | grater | [greitə(r)] | 擦子 |
| dish | [diʃ] | 深盘 | cutting **board** | [bɔ(r)d] | 切菜板 |
| fish platter | [fiʃ 'plætə(r)] | 鱼盘 | anti-hot **folder** | [fəuldə(r)] | 防烫夹 |
| skimmer | ['skimə(r)] | 漏勺 | colander | [kʌləndə(r)] | 沥水篮 |
| rolling pin | [rəuliŋ pin] | 擀面杖 | brush | [brʌʃ] | 刷子 |
| spatula | ['spætʃələ] | 铲子 | sieve | [siv] | 筛子 |
| mallet | [mælət] | 肉锤 | drain | [drein] | 沥干 |
| ladle | [leid(ə)l] | 炒勺 | slice | [slais] | 切片 |
| seasoning **pot** | [pɔt] | 调味罐 | pastry | ['peistri] | 点心 |
| soup spoon | [suːp spuːn] | 汤勺 | cookware | [kukweə(r)] | 炊具 |
| chopsticks | ['tʃɔpˌstiks] | 筷子 | **increase** the speed | [inkris] | 提高速度 |
| **pressure** cooker | ['preʃə(r)] | 压力锅 | commonly | [kɔmənli] | 通常 |
| **electric** cooker | [i'lektrik] | 电饭锅 | be **suitable** for | [suːtəb(ə)l] | 适合 |
| frying pan | ['fraiiŋ pæn] | 煎锅 | stew | [stjuː] | 炖 |
| **braising** pan | [breiziŋ] | 焖锅 | stir-fry | [stərfrai] | 炒菜 |

Unit 4  Tools & Utensils

## Learning Tips（学习提示）

　　食品卫生是非常重要的，厨师在使用菜板准备食材时，要注意生食和熟食分开，以免滋生细菌。此外，还要注意水果、蔬菜和鱼类分开，以免影响菜肴的味道。用不同颜色的菜板，分类使用更健康。

### 菜板的分类

　　菜板的出现使厨房卫生得到了改善。
　　传统意义上的菜板集切蔬菜、水果、鱼肉功能于一体，很不卫生。现代菜板主要分为4类：蓝色、红色、绿色和黄色，它们分别用于切鱼、切肉、切蔬菜和切水果，既美观、方便又卫生。蔬菜、水果、肉类、海鲜得到了区分，同时，避免了食材间味道的相互影响，是厨房的好帮手！

 ## Culture Knowledge（文化知识）

### 中国碗筷文化

　　中国碗筷不仅是食具，更是中国文化的传承。

　　筷子的原型叫箸，如李白《行路难》的"停杯投箸不能食"。准确地说，唐代对筷子的叫法是"筋"。到了宋代，筷子才被称为"筷"。筷子简单实用，体现了中国儒家思想不动刀枪、以和为贵的泰然处世之理。

　　筷子作为一件常用物品，存在许多禁忌和寓意。人们忌讳将两根筷子并排直立插在饭中，认为这是不吉利的象征。吃饭的时候不要不停在盘子周围转来转去，不停在盘子周围转来转去的行为会被看成没修养。

　　中国碗一般为陶瓷材质，有白瓷、花瓷、青瓷几种样式，并且加入了各种图案。

　　俗话说，一方水土养一方人。一件经久不衰的物品背后必然折射出一个民族的光辉文化。

# Unit 5
# Seasonings

## You will be able to:

- know the words on condiments;
- know how to use the seasoning words;
- talk about different tastes of dishes.

# Warming Up（热身）

## 1 Look and match（看图并连线）

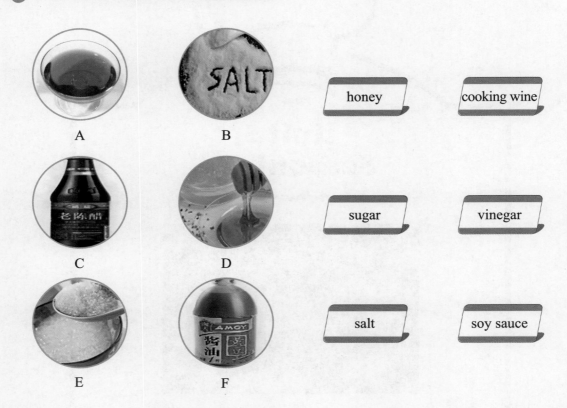

## 2 Decide true or false（判断正误）

1._____ Honey is widely used in cakes.

2._____ We need much salt when we make cakes.

3._____ Cooking wine is often used when we cook meat.

4._____ Vinegar should be used in sweet and sour food.

5._____ Many people in different areas like to add sugar to cold dishes.

# Now I Can Learn（学一学）

## Part A

### 1 Read and write（朗读并写出各种油的英文表达）

salad oil 色拉油　　rapeseed oil 菜籽油　　sesame oil 香油　　mixed edible oil 食用调和油
lard oil 猪油　　　　peanut oil 花生油　　　olive oil 橄榄油　　oyster oil 蚝油

1. _____　　2. _____　　3. _____　　4. _____

5. _____　　6. _____　　7. _____　　8. _____

### 2 Drill（句型训练）

**A:** What would you like to make today?

**B:** I'd like to deep fry some peanuts.

**A:** Can you tell me how to control the temperature of oil?

**B:** We usually use the low heat when we deep fry the peanuts.

### 3 Listen, complete and read（听录音，完成并朗读对话）

**A** control the temperature of oil　　**B** what would you like to make

**C** can I use the lard　　**D** Absolutely the high heat

Unit 5　Seasonings

**Jack:** Can you tell me how to use the oil correctly?

**Chef:** Yes, of course. _____ today?

**Jack:** I'd like to deep fry some peanuts. _____?

**Chef:** No. You'd better use the peanut oil or mixed edible oil.

**Jack:** I don't know how to _____ . Can you tell me the skill?

**Chef:** We usually use the low heat when we deep fry the peanuts.

**Jack:** If we deep fry ducks?

**Chef:** _____ .

**Jack:** I see. Thank you.

**Notes:**
absolutely 绝对地

## Part B

### 1 Listen and read（听录音，读单词和词组）

| wild pepper 花椒 | mustard 芥末 | chicken powder 鸡精 | pepper 胡椒 | fennel 小茴香 |
| cornstarch 玉米淀粉 | MSG 味精 | cinnamon 桂皮 | aniseed 大料 | bay leaf 香叶 |

### 2 Look, write and talk（看图写英文，并进行对话练习）

1. _____
2. _____
3. _____
4. _____
5. _____
6. _____
7. _____
8. _____
9. _____
10. _____

**A:** What kind of seasonings do you often use during cooking?

**B:** We often use _____ .

56 中餐烹饪英语

## 3 Read（读重点词汇）

1. be widely used 被广泛使用
2. pickled food 卤菜
3. sprinkle some sesame oil on it 在上面淋些香油

## 4 Choose and read（选择正确答案，并朗读对话）

**Jack:** Could you tell me how to use cornstarch and mustard, chef?
**Chef:** Cornstarch _____ , especially when we cook fish and meat.
**Jack:** How about aniseed, fennel, cinnamon and bay leaf?
**Chef:** They are often used _____ .
**Jack:** Is it important to _____ and chili to make spicy food?
**Chef:** Yes. Especially Sichuan food. It's more delicious to _____ .
**Jack:** Thanks a lot. I got it.

A. add some wild pepper
B. sprinkle some sesame oil
C. is widely used in hot dishes
D. in pickled food

## Part C

### 1 Read and tick（读单词和词组，勾出你常用的调料）

ketchup 番茄酱　　sweet sauce 甜面酱　　fish sauce 鱼露　　soy bean paste 豆瓣酱
rock sugar 冰糖　　ginger sauce 姜汁　　chili oil 辣椒油　　rice wine 黄酒

## 2 Match（连线）

1. salty      A
2. hot / spicy      B
3. sweet      C
4. bitter      D
5. sour      E

## 3 Fill in and evaluate（补全单词，并进行自我评价）

1. s_gar    2. s_same    3. s_lt    4. h_t    5. v_negar

6. s_y sauce    7. l__d    8. ch_l_    9. p_pper    10. sw__t

Assessment: If you can write 8-10 words, you are perfect.
               If you can write 4-7 words, you are good.
               If you can write 1-3 words, you need to try again.

# Now I Can Speak (说一说)

## Make dialogues (对话练习)

**e.g.**
**A:** Do you have any *soy sauce*?
**B:** Yes. How much do you want?
**A:** A *tablespoon*.
**B:** Here you are.
**A:** Thank you.

sugar / spoon
cornstarch / bowl
sesame oil / bottle
MSG / bag

**e.g.**
**A:** Could you tell me how to call it in English?
**B:** It is called *sugar*.
**A:** What's the taste of it?
**B:** It's *sweet*.

chili oil / hot or spicy     vinegar / sour     black coffee / bitter     salt / salty

Unit 5  Seasonings

## Now I Can Read（读一读）

**1** Read and answer（读对话，回答问题）

**Jack:** Could you tell me how to make Chongqing hot pot?
**Chef:** Sure. But it's a little difficult.
**Jack:** Really? Why?
**Chef:** We need more than 20 condiments to make a mixture.
**Jack:** So what are they?
**Chef:** The main condiments are red dry pepper, ginger, wild pepper, salt, garlic, Pixian soy bean paste, rapeseed oil, chili sauce and so on.
**Jack:** Can you tell me the way to make this dish?
**Chef:** Yes. First boil the red dry pepper in a pot and mince it all.
**Jack:** And then?
**Chef:** Then put salad oil and the minced chili into the pot, add the seasonings left in the pan. Stir-fry the mixture for over two hours.
**Jack:** That's very interesting. Let me have a try.

1. What will Jack make today?

   _____

2. What condiments are needed for Hot Pot?

   _____

3. Shall I fry the red dry pepper? What should I do?

   _____

4. How long should we stir-fry the mixture?

   _____

## 2 Write and tick (写出英文表达，勾出重庆火锅所需要的调料)

1._____  2._____  3._____  4._____

5._____  6._____  7._____  8._____

## 3 Write (根据图片写出制作过程)

**1** _____ the red dry pepper and _____.

**2** Put in the _____.

**3**

**4**

Unit 5  Seasonings  61

# Now I Can Do（练一练）

**Exercise 1**  Translate into Chinese or English（单词和词组中英文互译）

1. olive oil _____
2. 盐 _____
3. rapeseed oil _____
4. 辣椒油 _____
5. aniseed _____
6. 豆瓣 _____
7. soy sauce _____
8. 花生油 _____
9. sweet and sour _____
10. 冰糖 _____

**Exercise 2**  Tick off the odd words（勾出不同类别的单词或词组）

1. A. vinegar      B. meat        C. pepper      D. soy sauce
2. A. tomato      B. aniseed     C. honey       D. cooking wine
3. A. salt         B. sugar       C. vinegar     D. sliced fish
4. A. sweet sauce  B. cinnamon    C. beef        D. chili oil
5. A. coffee       B. mustard     C. bay leaf    D. cornstarch

**Exercise 3**  Fill in the blanks（填空）

1. People can't live without s_____（盐）.

2. We can't put M_____（味精）into the boiling soup.

3. L_____（猪油）is from the pig's fat.

4. W_____ pepper（花椒）is widely used in Sichuan food.

5. People like to add some v_____（醋）to the cold dishes.

6. A_____（大料）is used for pickled food（卤菜）.

7. P_____ o_____（花生油）comes from peanut.

8. C_____（辣椒）means dry red pepper.

9. We often put some s_____ oil（香油）in the dipping sauce when we have hot pot.

10. Seafood is often served with m_____（芥末）.

*Exercise 4*  **Find the answers**（找答案）

A. Could you pass me the chicken powder?
B. Do you have any salad oil?
C. How much do you want?
D. How can I use the sesame oil?
E. Can you tell me how to make the sauce?

1. Mix the salt, vinegar, MSG, cornstarch, sugar, minced ginger and so on.
2. Yes, of course.
3. A tablespoon, please.
4. Sprinkle it on the duck.
5. Sure. Here you are.

*Exercise 5*  **Choose the best answer**（选词填空）

sour    sweet    salty    hot    sweet and sour

1. Cold peanuts with vinegar is a little _____ .
2. Some people add some honey to the cake to make it more _____ .
3. Some seafood can be served without salt, because they are _____ .
4. Diced pork with chili is very _____ .
5. Fish with sugar and vinegar is _____ .

*Exercise 6*  **Translate into Chinese or English**（句子中英文互译）

1. 我们做油炸花生时通常使用小火。
2. 汤里加点胡椒粉，这样会更加鲜美。
3. 这道凉菜要用葱、姜、蒜、酱油、味精、醋及白糖调和而成。
4. Pass me a glass of cooking wine, please.
5. Boil the red dry pepper in a pot and mince it all.

# Word List（单词表）

| | | | | | | |
|---|---|---|---|---|---|---|
| honey | ['hʌni] | 蜂蜜 | aniseed | ['ænəsi:d] | 大料 |
| cooking wine | — | 料酒 | bay leaf | [bei li:f] | 香叶 |
| sugar | ['ʃugə(r)] | 白糖 | ketchup | ['ketʃəp] | 番茄酱 |
| vinegar | ['vinigə(r)] | 醋 | sweet sauce | [swi:t sɔ:s] | 甜面酱 |
| salt | [sɔ:lt] | 盐 | fish sauce | [fiʃ sɔ:s] | 鱼露 |
| soy sauce | [sɔi 'sɔ:s] | 酱油 | soy bean paste | ['sɔi bi:n peist] | 豆瓣酱 |
| salad oil | ['sæləd ɔil] | 色拉油 | rock sugar | [rɔk 'ʃugə(r)] | 冰糖 |
| **rapeseed** oil | ['reipsi:d] | 菜籽油 | **ginger** sauce | ['dʒindʒə(r)] | 姜汁 |
| **sesame** oil | ['sesəmi] | 香油 | **chili** oil | ['tʃili] | 辣椒油 |
| mixed **edible** oil | ['edəbl] | 食用调和油 | rice wine | — | 黄酒 |
| lard | [lɑ:d] | 猪油 | salty | ['sɔ:lti] | 咸的 |
| **peanut** oil | ['pi:nʌt] | 花生油 | hot /**spicy** | ['spaisi] | 辛辣的 |
| **olive** oil | ['ɔliv] | 橄榄油 | sweet | [swi:t] | 甜的 |
| **oyster** oil | ['ɔistə(r)] | 蚝油 | bitter | ['bitə(r)] | 苦的 |
| wild pepper | [waild pepə(r)] | 花椒 | sour | ['sauə(r)] | 酸的 |
| mustard | ['mʌstəd] | 芥末 | absolutely | ['æbsəlu:tli] | 绝对地 |
| chicken **powder** | ['paudə(r)] | 鸡精 | widely | ['waidli] | 广泛地 |
| pepper | ['pepə(r)] | 胡椒 | pickled | ['pikld] | 腌渍的 |
| fennel | ['fenl] | 小茴香 | sprinkle | ['spriŋkl] | 把……洒（撒）在…… |
| cornstarch | ['kɔ:nstɑ:tʃ] | 玉米淀粉 | tablespoon | ['teiblspu:n] | 餐匙 |
| MSG | — | 味精 | condiment | ['kɔndimənt] | 调味料 |
| cinnamon | ['sinəmən] | 桂皮 | | | |

# Learning Tips（学习提示）

　　中餐的调料极其丰富。除常见的油、盐、酱、醋等调料外，很多中药材、蔬菜和果汁都用作调料，形成了各式各样的口味，以满足人们不同的饮食需求。以川菜为例，常见川菜的有 24 种味，如下表所示。

| 1. 麻辣味 | 2. 酸辣味 | 3. 泡椒味 | 4. 怪味 | 5. 糊辣味 | 6. 红油味 |
| 7. 家常味 | 8. 鱼香味 | 9. 荔枝味 | 10. 咸鲜味 | 11. 甜香味 | 12. 烟香味 |
| 13. 椒麻味 | 14. 蒜泥味 | 15. 五香味 | 16. 糖醋味 | 17. 咸甜味 | 18. 陈皮味 |
| 19. 酱香味 | 20. 姜汁味 | 21. 麻酱味 | 22. 椒盐味 | 23. 香糟味 | 24. 芥末味 |

# Culture Knowledge（文化知识）

## 中国各地区口味

　　因为地理位置和传统习惯的不同，每个地方都有自己主打的味道。

　　因为东北地区所处的地理位置纬度较高，一年之间气候寒冷的天气比较多，所以这里的人喜欢酸甜中带辣的味道。比如东北冷面，就是酸甜辣口的。

　　生活在西北地区的人们饮食古朴粗犷，爱吃肉，也喜辣味。

　　生活在华北地区的人们以面食为主，口味偏重，酱油是不可缺少的原料之一。

　　在西南地区，有"宁可无菜，不可缺辣"的说法。这里的人们也喜酸，常说"三日不吃酸，走路打转转"。

　　生活在华南地区的人们有喝早茶和夜宵的习惯，一日五食，颇为讲究。

# Unit 6
# Fruits & Nuts

## *You will be able to:*

◇ get familiar with fruits and nuts;
◇ know about the basic sentences for fruits;
◇ talk about the ways of making fruit dishes.

# Warming Up(热身)

**1** Look and match(看图并连线)

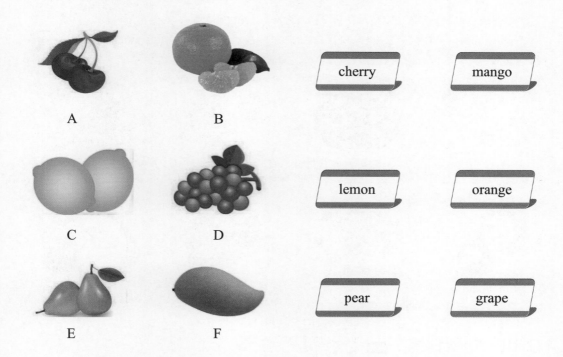

**2** Write(写出水果的英文表达)

1._____ It's yellow and sour, which is always used to go with the black tea.

2._____ It's a yellow juicy fruit, narrow at one end and wide at the other.

3._____ It's purple and grows in bunches(一串)on vines(藤).

4._____ A small round red fruit that starts with a "C".

5._____ It's a fruit with yellow or red skin, soft and sweet, which has a large long hard seed inside.

# Now I Can Learn（学一学）

## Part A

**1** Read and write（朗读并写出下列水果的英文表达）

> pineapple 菠萝　　peach 桃　　lychee 荔枝　　kiwi fruit 猕猴桃
> dragon fruit 火龙果　　strawberry 草莓　　coconut 椰子　　papaya 木瓜

1._____　2._____　3._____　4._____

5._____　6._____　7._____　8._____

**2** Drill（句型训练）

**A** What do we call these fruits in English?

**B** They are called peach, pineapple and lychee.

**A** What shall I do with them?

**B** Wash and peel them, remove the seeds, and then cut them in half.

**3** Listen, complete and read（听录音，完成并朗读对话）

**A** Remove the seeds

**B** What shall we do with the mangoes

**C** Wash and peel them

**D** We'll make mango juice

68　中餐烹饪英语

**Jack:** _____?
**Chef:** _____ today.
**Jack:** What should I do first?
**Chef:** _____.
**Jack:** OK. It's finished. And then?
**Chef:** _____, please.
**Jack:** I see. It's done.
**Chef:** Put them in the liquidizer.

# Part B

## 1 Listen and read（听录音，读单词和词组）

| melon 甜瓜 | blueberry 蓝莓 | grapefruit 西柚 | hawthorn 山楂 |
| watermelon 西瓜 | mulberry 桑葚 | apricot 杏 | plum 李子 |

## 2 Look, write and talk（看图写英文，并进行对话练习）

1. _____    2. _____    3. _____

A : What fruit would you like to eat?

B : I'd like to eat _____ .

4. _____                                          8. _____

5. _____    6. _____    7. _____

Unit 6  Fruits & Nuts

## 3 Read（读重点词汇）

1. a cup of yogurt 一杯酸奶
2. mix up 搅拌
3. put…in the mixing bowl 把……放入拌菜碗里

## 4 Choose and read（选择正确答案，并朗读对话）

**Jack:** What are we going to do today?
**Chef:** _____ . Prepare one orange, two bananas and an apple.
**Jack:** Oh, I see. After that?
**Chef:** _____ , then put the fruits in a mixing bowl.
**Jack:** OK, It's done. _____ ?
**Chef:** Put two teaspoons of honey and _____ .
**Jack:** Mix it up?
**Chef:** That's right.

**A.** Peel them and cut them up
**B.** We'll make a fruit salad
**C.** a cup of yogurt in the salad
**D.** And then

## Part C

### 1 Read and tick（读单词和词组，勾出你喜欢吃的坚果）

| walnut 核桃 | peanut 花生 | chestnut 栗子 | cashew 腰果 |
| raisin 葡萄干 | Chinese date 枣 | wolfberry 枸杞 | longan 龙眼 |

## 2 Match（连线）

1. make a salad      A

2. wash the apples      B

3. soak the plums      C

4. remove seeds      D

5. peel the bananas      E

## 3 Fill in and evaluate（补全单词，并自我评价）

1. p_ne_pple    2. str_wb_rry    3. l_ch_e    4. papay_    5. p_ar

6. waterm_lon    7. p_ _ch    8. pean_t    9. m_ngo    10. apric_t

Assessment: If you can write 8–10 words, you are perfect.
             If you can write 4–7 words, you are good.
             If you can write 1–3 words, you need to try again.

# Now I Can Speak（说一说）

## Make dialogues（对话练习）

**A:** What are you doing?
**B:** I'm _____ .
**A:** For what?
**B:** _____ .

**A:** What's the color of plums?
**B:** They are _____ .
**A:** Do you like them?
**B:** _____ .

**A:** What are these?
**B:** They're _____ .
**A:** Are they sweet or sour?
**B:** _____ .

**A:** Do you like watermelon?
**B:** _____ .
**A:** In which season do we have it?
**B:** _____ .

 eg.
**A:** Can you tell me which fruit is sour and which is sweet?
**B:** I think _____ is sour and _____ is sweet.

*Now I Can Read*（读一读）

**1** **Read and answer**（读对话，回答问题）

**Chef:** We are going to make banana milkshake（奶昔）today.
**Jack:** What shall we do first, chef?
**Chef:** Wash and peel two bananas, and then cut them in half, please.
**Jack:** Put them into the liquidizer?
**Chef:** Yes, then add some milk.
**Jack:** How much milk should I add?
**Chef:** Two glasses. Then stir them for a few seconds.
**Jack:** Oh, I got it. Sounds interesting.

1. What will they do with the bananas?
   _____

2. How long does it take to make banana milkshake with a liquidizer?
   _____

**2** **Write**（参考上一题，用英文写出香蕉奶昔的制作过程）

1._____    2._____    3._____

**3** **Tick**（勾出香蕉奶昔需要的原料）

☐ honey      ☐ banana      ☐ water      ☐ milk
☐ liquidizer  ☐ egg         ☐ sugar      ☐ juice

## 4 Read and decide（读菜谱，判读正误）

### Fruits Congee（水果粥）
### Ingredients

1 apple    1 pear    1 orange    1 banana
raisins    rock sugar（冰糖）    30g lotus root starch（藕粉）

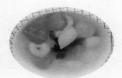

### Directions

1. Wash the fruits.
2. Peel and cut them into small pieces.
3. Put the lotus root starch into a small bowl, add some cold water, and then stir it.
4. Put the fruits into the pot. When the water is boiling, add some rock sugar.
5. Add the lotus root starch into the pot and stir the congee for 1 minute.
6. At last, turn off the heat and drop some raisins.

1._____ Cut the fruits into small pieces when you cook the fruits congee.

2._____ Add some hot water into the lotus root starch.

3._____ Boil the fruits for about 10 minutes.

4._____ Add some honey to the pot.

5._____ Drop some raisins when you turn off the heat.

## 5 Translate and write（菜品原料中英文互译，并写出加工流程的英文表达）

### Ingredients

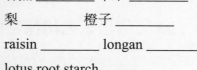

香蕉_____ 苹果_____
梨_____ 橙子_____
raisin_____ longan_____
lotus root starch_____
rock sugar_____

## Production process

**1** Wash the _____.

**2** _____ and _____ them into small pieces.

**3** Add some cold water into the _____.

**4**

**5**

**6**

## Now I Can Do（练一练）

**Exercise 1** Translate into Chinese or English（单词和词组中英文互译）

1. peanut _____
2. 菠萝 _____
3. Chinese date _____
4. 木瓜 _____
5. chestnut _____
6. 荔枝 _____
7. walnut _____
8. 枸杞 _____
9. longan _____
10. 杧果 _____

Unit 6　Fruits & Nuts　75

*Exercise 2*  **Tick off the odd words**（勾出不同类别的单词或词组）

1. A. peach	B. apple	C. banana	D. peanut
2. A. chestnut	B. apricot	C. grape	D. peach
3. A. pineapple	B. cashew	C. melon	D. cherry
4. A. mango	B. papaya	C. raisin	D. kiwi fruit

*Exercise 3*  **Find the answers**（找答案）

1. What are you doing?
2. What's the color of the grape?
3. Do you like plum?
4. What shall we do with the melon?
5. What about the next?

A. Put them in the liquidizer.
B. Just so so, it's too sour.
C. I'm washing the fruits.
D. It's purple.
E. Cut it in half and remove seeds.

*Exercise 4*  **Make sentences in right order**（连词成句）

1. the, fruits, boil, about, minutes, for, ten

   _____

2. and, peel, the bananas, cut, and then, them, half, wash, in

   _____

3. plum, is, color, of, what, the

   _____

4. season, we, peach, in, have, which, do

   _____

5. it, seeds, of, and then, slice, remove, the, melon

   _____

## Exercise 5   Fill in the blanks（填空）

1. _____ a fruit salad.（做）
2. _____ the grapes.（洗）
3. _____ the pears.（去皮）
4. _____ the seeds from the melon.（去除）
5. Make _____.（香蕉奶昔）
6. Put the mango into the_____.（榨汁机）
7. The little girl likes _____ very much.（水果粥）
8. _____ the apples into _____.（切成小块）

## Exercise 6   Translate into Chinese or English（句子中英文互译）

1. 今天我们要做草莓奶昔。
2. 请把枣洗了。
3. 腰果和栗子，你更喜欢吃哪一个？
4. Add some cold water into the lotus root starch, and then stir it.
5. Drop some wolfberries into the soup.

 # Word List（单词表）

| pineapple | ['painæpl] | 菠萝 | cashew | ['kæʃu:] | 腰果 |
| peach | [pi:tʃ] | 桃 | raisin | ['reizn] | 葡萄干 |
| lychee | [ˌlai'tʃi:] | 荔枝 | Chinese **date** | [deit] | 枣 |
| kiwi fruit | ['ki:wi: fru:t] | 猕猴桃 | wolfberry | ['wulfberi] | 枸杞 |
| **dragon** fruit | ['drægən] | 火龙果 | longan | ['lɔŋgən] | 龙眼 |
| strawberry | ['strɔ:bəri] | 草莓 | a cup of **yogurt** | ['jɔgət] | 一杯酸奶 |
| coconut | ['kəukənʌt] | 椰子 | mix up | ['miks ʌp] | 搅拌 |
| papaya | [pə'paiə] | 木瓜 | put... in | — | 把……放入 |
| melon | ['melən] | 甜瓜 | bunch | [bʌntʃ] | 串，束 |
| blueberry | ['blu:bəri] | 蓝莓 | vine | [vain] | 藤 |

续表

| grapefruit | ['greipfru:t] | 西柚 | seed | [si:d] | 种子，果核 |
|---|---|---|---|---|---|
| hawthorn | ['hɔ:θɔ:n] | 山楂 | teaspoon | ['ti:ˌspu:n] | 茶匙 |
| watermelon | ['wɔ:təmelən] | 西瓜 | soak | [səuk] | 浸泡 |
| mulberry | ['mʌlbəri] | 桑葚 | milkshake | ['milkʃeik] | 奶昔 |
| **apricot** | [eiprikɔt] | 杏 | congee | ['kɔndʒi:] | 粥 |
| plum | [plʌm] | 李子 | ingredient | [in'gri:diənt] | 原料 |
| walnut | ['wɔ:lnʌt] | 核桃 | lotus root | ['ləutəs ru:t] | 莲藕 |
| peanut | ['pi:nʌt] | 花生 | lotus root **starch** | [stɑ:tʃ] | 藕粉 |
| chestnut | ['tʃesnʌt] | 栗子 | | | |

# *Learning Tips*（学习提示）

新鲜水果汁含有丰富的维生素和微量元素，对皮肤和身体极为有益，其品质的纯净性与天然性，是任何昂贵的护肤品和保健品都无法超越的。

| 红色恋人 | ◆ 苹果 / 橙 / 胡萝卜 | 消除疲劳 |
| 红粉佳人 | ◆ 西红柿 / 橙 / 菠萝 / 柠檬 | 预防皮肤老化 |
| 挪威森林 | ◆ 奇异果 / 苹果 / 菠萝 | 排毒养颜 |
| 浪漫宣言 | ◆ 西红柿 / 橙 | 预防高血压，延缓衰老 |
| 瘦身汁 | ◆ 黄瓜 / 菠萝 | 减肥，预防口腔炎症，舒缓喉痛 |
| 润肤汁 | ◆ 西红柿 / 橙 / 杨桃 | 预防皮肤干燥及头皮屑过多 |
| 化痰汁 | ◆ 杨桃 / 菠萝 | 改善浓痰咳嗽 |
| 润肺汁 | ◆ 杨桃 / 菠萝 / 胡萝卜 | 润肺，适合长期吸烟者 |
| 维 C 果汁 | ◆ 苹果 / 橙 | 促进新陈代谢，净化肠道 |
| 抗氧化汁 | ◆ 胡萝卜 / 橙 | 增强抗氧化能力 |

# Culture Knowledge（文化知识）

近年来，坚果丰富的营养已经被越来越多的人认可。坚果种类繁多，其功效和作用也各有不同，大家可以根据自己的需求进行选择。

核桃含有丰富的蛋白质、脂肪、锌、锰、B族维生素和维生素E。由于锌和锰是脑垂体的重要成分，因此核桃是健脑益智的营养佳品。

腰果含有丰富的蛋白质、脂肪、淀粉、多种维生素，以及磷、硒、镁等多种矿物质。腰果中的脂肪主要为亚麻酸和不饱和脂肪酸等，对降低胆固醇、预防心脑血管疾病有很好的疗效。

花生中含有大量的蛋白质和脂肪，还含有胆碱、B族维生素、维生素K、维生素A、维生素E、硒、钙等20多种微营养素。其中，硒具有防治肿瘤、预防动脉粥样硬化和心脑血管疾病的作用。

栗子含有丰富的蛋白质、脂肪、维生素C、B族维生素、膳食纤维，以及钙、磷、铁、钾等多种矿物质。栗子能防治高血压、冠心病、动脉粥样硬化、骨质疏松等疾病。栗子是抗衰老、延年益寿的滋补食品。

# Unit 7
# Vegetables

## You will be able to:

◇ get familiar with different vegetables;
◇ describe the vegetable preparation process;
◇ talk about the process of making vegetable dishes.

# Warming Up（热身）

**1** Look and match（看图并连线）

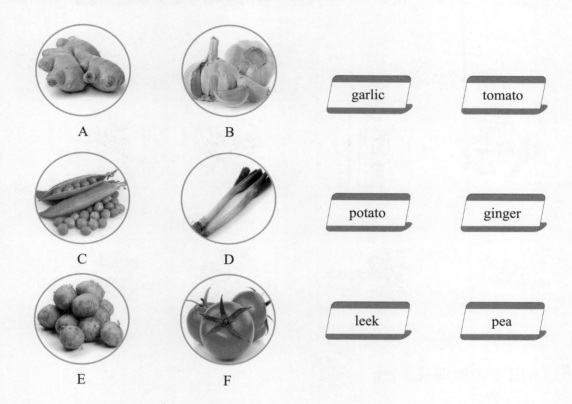

**2** Decide true or false（判断正误）

1. _____ Leeks are green and white and can be used in the porridge.

2. _____ Tomatoes are red and round.

3. _____ Potatoes are yellow, thin and long.

4. _____ Peas are white and thin.

5. _____ French beans must be cooked with garlic.

Unit 7   Vegetables   81

# Now I Can Learn（学一学）

## Part A

**1** Read and write（读单词和词组，写出蔬菜的英文表达）

| asparagus 芦笋 | lettuce 生菜 | celery 芹菜 | spinach 菠菜 |
| Chinese cabbage 大白菜 | parsley 香芹 | crown daisy 茼蒿 | broccoli 西兰花 |

1._____ 2._____ 3._____ 4._____

5._____ 6._____ 7._____ 8._____

**2** Drill（句型训练）

A: How do you cut asparagus and celery?
B: I cut sliced asparagus and chopped celery.

A: What would you like me to do?
B: I'd like you to cut the spinach into batons and slice Chinese cabbages.

**3** Listen, complete and read.（听录音，完成并朗读对话）

A: What shall I do first
B: cut sliced onions and chopped parsley
C: And what about the parsley
D: Chop the parsley finely

82　中餐烹饪英语

**Jack:** What shall we cook today?
**Chef:** We will make a new vegetable salad.
**Jack:** What are we going to do?
**Chef:** We are going to _____.
**Jack:** _____?
**Chef:** Wash the onions and slice them.
**Jack:** _____?
**Chef:** _____.
**Jack:** All right, I see.

## Part B

### 1 Listen and read（听录音，读单词和词组）

| green pepper 青椒 | white gourd 冬瓜 | cucumber 黄瓜 | coriander 香菜 |
| bean 青豆 | eggplant 茄子 | mung bean 绿豆 | bean sprout 豆芽 |

### 2 Look, write and talk（看图写英文，并进行对话练习）

1. _____

2. _____

3. _____

4. _____

A: What are you cooking?
B: I'm cooking _____.

8. _____

5. _____

6. _____

7. _____

## 3 Read（读重点词汇）

1. put ... in the mixing bowl 把……放入拌菜碗里
2. shred the green pepper 把青椒切成丝
3. chop the coriander finely 把香菜切成末

## 4 Choose and read（选择正确答案，并朗读对话）

Jack: What are you going to do?
Chef: I'm going to _____.
Jack: What do you want me to do?
Chef: Help me to _____.
Jack: What shall I do with the coriander?
Chef: You can _____.
Jack: Oh, I see. Now it's done.
Chef: Now _____.

A. put them in a mixing bowl
B. chop the coriander finely
C. shred the green pepper
D. wash the green pepper

# Part C

## 1 Read and tick（读单词和词组，勾出你喜欢吃的蔬菜）

☐ French bean 四季豆　☐ pumpkin 南瓜　☐ soy bean 黄豆　☐ kidney bean 芸豆
☐ onion 洋葱　☐ broccoli 西兰花　☐ black fungus 木耳　☐ snow pea 荷兰豆

84　中餐烹饪英语

## 2 Match（连线）

1. dice the onion — A

2. cut off the bottom of the onion — B

3. cut the onion in half — C

4. slice the onion — D

## 3 Fill in and evaluate（补全单词，并自我评价）

1. _nion    2. p_mpkin    3. l_tt_ce    4. br_ccoli    5. c_bbage

6. sp_nach    7. c__cumb_r    8. _sparagus    9. d_ce    10. c_l_ry

Assessment: If you can write 8–10 words, you are perfect.
If you can write 4–7 words, you are good.
If you can write 1–3 words, you need to try again.

Unit 7  Vegetables  85

# Now I Can Speak（说一说）

## Make dialogues（对话练习）

**e.g.**
**A:** What are you doing?
**B:** I'm *chopping*.
**A:** Chopping what?
**B:** *Chopping the celery*.

peel / cucumber
grind / garlic
slice / carrot

**e.g.**
**A:** What kind of vegetable dish can you cook?
**B:** I can cook _____.
**A:** How do we call it in Chinese?
**B:** It is called _____.

bitter melon in sauce
凉拌苦瓜

tiger salad
老虎菜

bean sprouts with green pepper
豆芽菜拌青椒

pea sprouts in sauce
拌豆苗

# Now I Can Read（读一读）

**1** Read and answer（读对话，回答问题）

Jack: Could you tell me how to make black fungus with onion, chef?
Chef: No problem. First, you should prepare black fungus, onion, green and red pepper.
Jack: I know that. Do we need soy sauce, salt and vinegar?
Chef: Yes, and also sugar, sesame oil and gourmet powder. Then cut sliced onion and black fungus. Boil the sliced onion for 2 minutes.
Jack: OK, how about ingredients?
Chef: Sliced green and red pepper, sweet sauce, vinegar, salt and sugar.
Jack: I've finished the work.
Chef: Let's do it now.

1. How do they prepare the vegetables?
   _____

2. What should be put in the dish as ingredients?
   _____

**2** Write（参考上一题，用英文写出菜品的制作过程）

  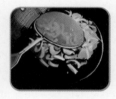  

1._____      2._____      3._____

Unit 7  Vegetables  87

## 3 Tick（勾出制作菜品需要的原料）

- ☐ sesame oil
- ☐ soy sauce
- ☐ green pepper
- ☐ vinegar
- ☐ salt
- ☐ sugar
- ☐ red pepper
- ☐ garlic

## 4 Read and decide（读菜谱，判读正误）

### Snow Peas with Grinded Garlic
### Ingredients

45 ml peanut oil  200 g snow peas  1/2 tbsp soy sauce
2 tbsp grinded garlic  1/4 tsp gourmet powder  1/4 tsp salt

### Directions
1. Boil the snow peas in boiling water for two minutes and dry them.
2. Stir-fry the grinded garlic in hot peanut oil for two minutes until heated through.
3. Put the snow peas into peanut oil, stir-fry for two minutes.
4. Add soy sauce, salt and gourmet powder, stir-fry for one minute.

### Characteristics
Snow peas with grinded garlic is crisp and fresh, with an attractive green color.

1._____ Boil snow peas in cold water for two minutes.

2._____ Stir-fry snow peas in peanut oil first.

3._____ Stir-fry grinded garlic in oil for one minute.

4._____ Snow peas with grinded garlic is sweet and green.

5._____ We don't need soy sauce while cooking this dish.

## 5 Translate and write（菜品原料中英文互译，并写出菜肴制作过程的英文表达）

### Ingredients

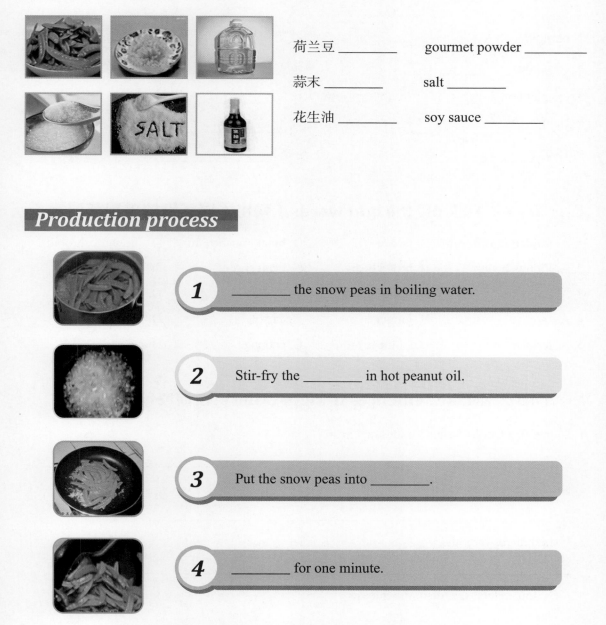

荷兰豆 _____     gourmet powder _____

蒜末 _____      salt _____

花生油 _____    soy sauce _____

### Production process

1. _____ the snow peas in boiling water.

2. Stir-fry the _____ in hot peanut oil.

3. Put the snow peas into _____.

4. _____ for one minute.

**Now I Can Do**（练一练）

### Exercise 1　Translate into Chinese or English（单词和词组中英文互译）

1. pumpkin _____
2. 西兰花 _____
3. coriander _____
4. 青椒 _____
5. eggplant _____
6. 大白菜 _____
7. soy bean _____
8. 茼蒿 _____
9. 芹菜 _____
10. 胡萝卜 _____

### Exercise 2　Tick off the odd words（勾出不同类别的单词或词组）

1. A. golden mushroom　　B. taro　　　　　　C. black mushroom　　D. straw mushroom
2. A. kidney bean　　　　　B. broad bean　　　C. bean sprouts　　　　D. parsley
3. A. pumpkin　　　　　　B. bitter melon　　　C. garlic　　　　　　　D. cucumber
4. A. taro　　　　　　　　B. ginger　　　　　 C. carrot　　　　　　　D. mushroom
5. A. lettuce　　　　　　　B. Chinese toon　　 C. cabbage　　　　　　D. celery

### Exercise 3　Make sentences in right order（连词成句）

1. is, the, most, cold dishes, of, onion, "heart"

   _____.

2. slicing, green pepper, are, now, you

   _____.

3. I, shall, dice, yam, the

   _____.

4. for, chop, celery, the, dish, the, please

   _____.

5. cook, I, shall, how, spinach, the

   _____.

*Exercise 4*  **Fill in the blanks（填空）**

> peel    grind    chopping    put…to    batons    sliced

1. _____ the potatoes firstly.
2. For the preparation work, _____ the garlic finely.
3. Please _____ some salt _____ the broccoli.
4. Would you cut some _____ onion for black fungus with onion?
5. A: Shall I chop the carrots?
   B: No, cut carrot _____ carefully.
6. A: What are you doing?
   B: I'm _____ the celery.

*Exercise 5*  **Translate into English（写出下列菜品的英文表达）**

> Bitter Melon Salad                    Spiced Kidney Beans
> Sweet and Sour Pickled Vegetables    Crispy Celery
> Sliced Turnip with Sauce              Black Mushrooms with Pine Nuts

酸甜泡菜_____    五香芸豆_____

水晶萝卜_____    爽口西芹_____

松仁香菇_____    冰心苦瓜_____

*Exercise 6*  **Translate into Chinese or English（句子中英文互译）**

1. 我们需要切得很细的姜丝。
2. 我现在切莴笋片好吗？
3. 他正在捣蒜泥。
4. Please wash the pumpkin, and then dice it.
5. Salt the cucumber for five minutes.

# Word List（单词表）

| | | | | | |
|---|---|---|---|---|---|
| asparagus | [əˈspærəɡəs] | 芦笋 | mung bean | [mʌŋ biːn] | 绿豆 |
| lettuce | [ˈletis] | 生菜 | bean sprout | [spraut] | 豆芽 |
| celery | [ˈseləri] | 芹菜 | mixing bowl | [bəul] | 拌菜碗 |
| spinach | [ˈspinitʃ] | 菠菜 | French bean | [biːn] | 四季豆 |
| Chinese cabbage | [ˈkæbidʒ] | 大白菜 | pumpkin | [ˈpʌmpkin] | 南瓜 |
| parsley | [ˈpɑːsli] | 香芹 | soy bean | [sɔi] | 黄豆 |
| crown daisy | [kraun ˈdeizi] | 茼蒿 | kidney bean | [ˈkidni] | 芸豆 |
| broccoli | [ˈbrɔkəli] | 西兰花 | onion | [ˈʌnjən] | 洋葱 |
| green pepper | [pepə] | 青椒 | black fungus | [ˈfʌŋɡəs] | 木耳 |
| white gourd | [ɡuəd] | 冬瓜 | snow pea | [snəu piː] | 荷兰豆 |
| cucumber | [ˈkjuːkʌmbə(r)] | 黄瓜 | ingredient | [inˈɡriːdiənt] | 原料 |
| coriander | [ˌkɔriˈændə(r)] | 香菜 | cut in half | [hɑːf] | 切成两半 |
| bean | [biːn] | 青豆 | red pepper | [pepə] | 辣椒 |
| eggplant | [ˈeɡplɑːnt] | 茄子 | cold dish | — | 凉菜 |

 # Learning Tips（学习提示）

　　世界癌症研究基金会分析表明，蔬菜和水果能降低患癌症的风险。我们食用蔬菜时可以选择深色的，如菠菜、胡萝卜、紫甘蓝和南瓜等。

1. 蔬菜最好吃应季的、新鲜的　2. 多吃色彩丰富的蔬菜

3. 多换食不同种类的蔬菜　　4. 水果不能替代蔬菜

5. 多食用不同种类的菌类　　6. 食谱中应"顿顿有蔬菜"

7. 烹调蔬菜时应减少盐用量　8. 多吃根茎类蔬菜

# *Culture Knowledge*（文化知识）

## 一年四季吃什么蔬菜？

孔子曰"不时不食"，意思是不是当季的东西不吃。时令蔬菜是指根据蔬菜生长特点，在自然环境条件下，通过人工栽培管理或野生，采收后新鲜上市的蔬菜。

### 春季
春季是春暖花开、万物复苏的季节，有很多既美味而又有营养的蔬菜，它们是餐桌上的美味佳肴。如韭菜、油菜、菠菜、莴苣、竹笋、包菜、香椿、小白菜、木耳、蒜薹、芦笋、西兰花等都是不错的选择。

### 夏季
夏季蔬菜水分多、纤维素丰富，可以解暑补水、清肠排毒。夏季天气炎热，建议吃一些既营养丰富又清热解暑的蔬菜，如丝瓜、苦瓜、佛手瓜、红薯叶、黄瓜、冬瓜、茄子、豇豆、四季豆、西红柿、洋葱、苋菜、空心菜等。

### 秋季
秋季是成熟和收获的季节，有很多味道鲜美的瓜果蔬菜纷纷上市，加上秋季气候干燥，多吃些清淡可口的蔬菜可以滋阴润燥。秋季常见的蔬菜有茭白、莲藕、胡萝卜、秋葵、辣椒、菱角、花菜等。

### 冬季
相较于其他季节，冬季绿叶蔬菜的选择会少一点，冬季常见的蔬菜有大白菜、萝卜、芥菜、南瓜、红薯、荸荠等。

不同种类的蔬菜营养成分也不一样，经常变换种类，尽可能做到多样化，才能充分摄取不同食物中不同的营养素。

# Unit 8
# Meat

## *You will be able to:*

✧ get familiar with the names of different meats;

✧ know of how to describe some basic knife-cutting skills and cooking skills;

✧ describe a cooking process by using knife-cutting skills and cooking skills.

# Warming Up（热身）

## ① Look and match（看图并连线）

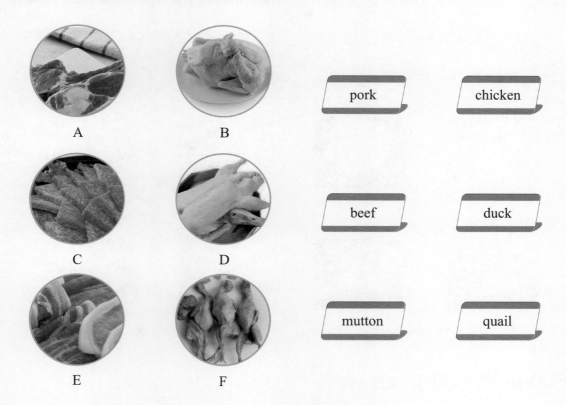

## ② Decide true or false（判断正误）

1. _____ Pork has more fat than beef and mutton.

2. _____ Chicken, duck and quail are three kinds of poultry（家禽）.

3. _____ We should eat meat as much as possible every day.

4. _____ People in some countries never eat pork.

5. _____ Fried food is delicious but eating too often or too much is harmful.

# Now I Can Learn（学一学）

## Part A

**1** Read and write（读单词和词组，并根据图片写出英文表达）

| tenderloin 里脊肉 | pork round 后臀尖 | pork belly 五花肉 | chicken thigh 鸡大腿 |
| shank 牛腱子 | lamb 羊腿 | spare ribs 排骨 | chicken breast 鸡脯肉 |

1._____  2._____  3._____  4._____

5._____  6._____  7._____  8._____

**2** Drill（句型训练）

**A** What are we going to learn?

**B** We're going to learn different kinds of meat.

**A** Are these meats used in different dishes?

**B** Yes, let's talk something about tenderloin, spare ribs, and chicken breast.

**3** Listen, complete and read（听录音，完成并朗读对话）

**A** How do we cook spare ribs

**B** We often shred it to make

**C** What about the chicken breast

**D** Kung Pao Chicken is famous

**Jack:** What shall we do with the pork tenderloin?
**Chef:** _____ Shredded Meat in Chili Sauce (鱼香肉丝).
**Jack:** Sounds great. _____ to cater the guests?
**Chef:** The guests like Sweet and Sour Spare Ribs very much.
**Jack:** Oh, I got it. _____ , chef?
**Chef:** We usually cut it into dice to make dishes, for example, _____ .

## Part B

### 1 Listen and read（听录音，读单词）

| cube 四方块 | chunk 不规则块 | dice 小丁 | mince 碎末，肉馅 |
| block 大块 | filet 鱼片，肉片 | roll 卷 | strip 条状 |

### 2 Look, write and talk（看图写英文，并进行对话练习）

1. _____    2. _____    3. _____

A: How do you call the shape of the meat?
B: It is called _____ .

4. _____

8. _____

5. _____    6. _____    7. _____

Unit 8  Meat

## 3 Read（读重点词汇）

1. Sweet and Sour Pork with Pineapple 菠萝咕噜肉
2. golden yellow 金黄色
3. season with 以……调味

## 4 Choose and read（选择正确答案，并朗读对话）

**Jack:** Could you tell me how to make _____?

**Chef:** Slice the tenderloin and _____.

**Jack:** What are we going to do next?

**Chef:** Cut a pineapple into chunks and _____.

**Jack:** And then what?

**Chef:** Fry the pork filet until golden yellow.

**Jack:** Is that all?

**Chef:** Stir-fry the ingredients for a short while and _____.

A. shred the green and red peppers

B. mix them with salt and egg white

C. Sweet and Sour Pork with Pineapple

D. season with ketchup and cooking wine

# Part C

## 1 Read and match（读单词或词组，匹配合适的图片）

☐ stir-fry 煸，爆炒    ☐ scramble 煎（鸡蛋）    ☐ scald 白灼，烫

☐ bake/roast 烘焙，烤    ☐ simmer/braise 煨，炖    ☐ grill （用烤架）烤

☐ deep fry 炸    ☐ steam 蒸

1._____    2._____    3._____    4._____

5._____  6._____  7._____  8._____

## 2 Match（连线）

1. cut the beef into cubes          A

2. cut the tenderloin into chunks   B

3. roll the sliced bacon belly      C

4. slice the shank                  D

5. mince the pork round             E

## 3 Fill in and evaluate（补全单词，并进行自我评价）

1. l_mb      2. sh_nk     3. st_ _m    4. scr_mble   5. sc_ld

6. s_mmer    7. r_ll      8. m_nce     9. c_be       10. sl_ce

Assessment: If you can write 8–10 words, you are perfect.

If you can write 4–7 words, you are good.

If you can write 1–3 words, you need to try again.

Unit 8  Meat

# Now I Can Speak（说一说）

## Make dialogues（对话练习）

**e.g.**
**A:** What shall I do with the *meat*?
**B:** *Cut* it into filet.
**A:** And then what?
**B:** *Grill* it for a while.

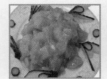  

chicken / dice / stir-fry
beef / block / braise
pork round / slice / steam

**e.g.**
**A:** What is this dish called in English?
**B:** It's called _____.
**A:** How do we usually cook it?
**B:** We usually _____ it.

   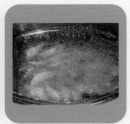

Steamed Pork with Rice Flour/steam
蒸米粉肉 / 蒸

Grilled Beef Filet/grill
扒牛柳/（用烤架）烤

Quick-fried Mutton Slices with Scallions/quick-fry
葱爆羊肉 / 爆炒

Soft-fried Pork Filet/fry
软炸肉片 / 炸

中餐烹饪英语

# Now I Can Read（读一读）

**1** Read and answer（读对话，回答问题）

Jack: Could you tell me some popular Chinese dishes?
Chef: Most of the foreigners like Pork Tenderloin with Sugar and Vinegar, Mapo Tofu , Beijing Roast Duck and so on.
Jack: I hear of "Kung Pao Chicken" is the most popular.
Chef: Yes. We also call it "Spicy Diced Chicken with Peanuts".
Jack: Could you tell me how to make it?
Chef: We need chicken thigh, fried peanuts, ginger, red chili … OK, now I'll show you how to make it.
Jack: That's wonderful.

1. Which dish is the most popular in the dialogue?
   _____
2. What ingredients should be prepared for the dish?
   _____

**2** Write（参考上一题，用英文写出菜名）

1._____  2._____  3._____  4._____

## 3 Tick（勾出制作宫保鸡丁需要的原料）

- ☐ starch
- ☐ peanut
- ☐ soy sauce
- ☐ scallion
- ☐ chili
- ☐ cucumber
- ☐ watercress
- ☐ ginger

## 4 Read and decide（读菜谱，判读正误）

### Spicy Diced Chicken with Peanuts
### Ingredients

300 g chicken breast, 50 g fried peanuts, 30 g chopped scallion, 20 g sliced ginger, 5 g red chili, an egg white, a little salt, soy sauce, starch and cooking wine.

### Directions

1. Flatten and dice the chicken breast.
2. Marinate the diced chicken with salt, egg white and starch for a while.
3. Fry the diced chicken until medium well.
4. Fry the red chili, ginger, scallion, and then stir-fry them with the diced chicken.
5. Put in the fried peanuts, a little soy sauce and cooking wine.

### Characteristics

The chicken is tender while the peanuts are crispy, which is spicy and tasty.

1. _____ Spicy Diced Chicken with Peanuts is welcomed by foreigners.
2. _____ We should fry the diced chicken for a very long time .
3. _____ Peanuts are necessary to make the dish crispy.
4. _____ Both ginger and scallion make the dish very spicy.
5. _____ The diced chicken is tough to eat.

*Notes*

肉类加热到几成熟的英文表达：
三成熟：medium rare
五成熟：medium
七成熟：medium well
全熟：well done

## 5 Translate and write（菜品原料中英文互译，并写出宫保鸡丁制作流程的英文表达）

### Ingredients

fresh grade breast _____   花生 _____
scallion _____   黄瓜 _____
ginger _____   蛋白 _____
cooking wine _____   辣椒 _____
watercress _____   淀粉 _____

### Production process

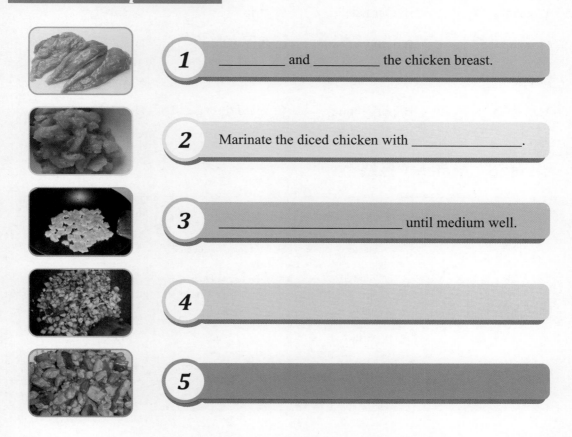

1. _____ and _____ the chicken breast.

2. Marinate the diced chicken with _____.

3. _____ until medium well.

4. 

5. 

Unit 8　Meat

# Now I Can Do（练一练）

### Exercise 1　Translate into Chinese or English（单词和词组中英文互译）

1. scramble _____
2. 片 _____
3. cube _____
4. 炸 _____
5. pork belly _____
6. 煨，炖 _____
7. spicy _____
8. 切碎的末 _____
9. tenderloin _____
10. 爆炒 _____

### Exercise 2　Tick off the odd words（勾出不同类别的单词或词组）

1. A. slice　　　　B. simmer　　　　C. shred　　　　D. cube
2. A. steam　　　　B. marinate　　　　C. strip　　　　D. stir-fry
3. A. steam　　　　B. boil　　　　C. deep fry　　　　D. mince
4. A. sugar　　　　B. cooking wine　　　　C. starch　　　　D. ginger
5. A. chunk　　　　B. pork round　　　　C. spare ribs　　　　D. chicken thigh

### Exercise 3　Match（连线）

| 菜名 | 烹饪技法 + ed | （形状）+ 主料 | with / in 辅料 |
|---|---|---|---|
| 架烤孜然羊肉 | Steamed | Sliced Chicken | with Ham |
| 香辣炸鸡腿 | Stewed | Diced Duck | in Broth（高汤） |
| 清蒸火腿鸡片 | Smoked | Spare Ribs | with Honey |
| 浓汤鸭块 | Grilled | Lamb | with Cumin（孜然） |
| 烟熏蜜汁肋排 | Fried | Chicken Thigh | with Chili |
| 爆炒回锅肉片 | Stir-fried | Sliced Pork | with Hot Spicy Sauce |

## Exercise 4　Choose the best answers（选择正确答案）

1. What_____ we going to learn?
   A. shall　　　　　B. are　　　　　　C. is　　　　　　　D. am
2. There are different ways of cutting, _____ there?
   A. isn't　　　　　B. aren't　　　　　C. don't　　　　　D. doesn't
3. Cut the tenderloin _____ filets and mix them _____ salt and egg white.
   A. into… with　　B. with… into　　C. from… in　　　D. from… to
4. A: Look! I'll show you _____ to make it.
   B: That's wonderful.
   A. what　　　　　B. where　　　　　C. how　　　　　　D. when
5. I know "Kung Pao Chicken" is _____ than the other dishes.
   A. more popular　B. the most popular　C. the more popular　D. most popular

## Exercise 5　Complete the sentences（补全句子）

e.g.　What shall I do with the beef?
　　　*What* shall I *do* with the chicken?

1. Why not cook the diced chicken for a long time?
   _____ _____ ask Beijing Roast Duck for lunch?
2. Could you tell me how to cut beef?
   _____ you _____ me how to make Kung Pao Chicken?
3. Both chili and watercress make the dish very spicy.
   _____ she _____ I like Sweet and Sour Pork with Pineapple.
4. I'll show you how to scramble the eggs.
   I'll _____ you _____ to stir-fry the tenderloin.

## Exercise 6　Translate into Chinese or English（句子中英文互译）

1. 我该怎样烹制宫保鸡丁？

2. 我们要把猪肉切成3种形状：肉丁、肉片和肉丝。

3. 顾客们通常喜欢点五香牛肉。

4. The chicken is tender while the peanuts are crispy.

5. I'm not sure how to say the Chinese dish in English.

# Word List（单词表）

| | | | | | |
|---|---|---|---|---|---|
| pork | [pɔːk] | 猪肉 | slice | [slais] | 把……切成（薄）片 |
| chicken | ['tʃikin] | 鸡肉 | scramble | ['skræmbl] | 煎（鸡蛋） |
| beef | [biːf] | 牛肉 | scald | [skɔːld] | 白灼，烫 |
| duck | [dʌk] | 鸭肉 | bake/roast | [beik] [rəʊst] | 焙烘，烤 |
| mutton | ['mʌtn] | 羊肉 | stir-fry | ['stəː frai] | 煸，爆炒 |
| simmer/braise | ['simə] [breiz] | 煨，炖 | grill | [gril] | （用烤架）烤 |
| quail | [kweil] | 鹌鹑 | deep fry | ['diːp frai] | 炸 |
| tenderloin | ['tendəlɔin] | 里脊肉 | steam | [stiːm] | 蒸 |
| pork round | ['raund] | 后臀尖 | vinegar | ['vinigə] | 醋 |
| pork belly | ['beli] | 五花肉 | peanut | ['piːnʌt] | 花生 |
| chicken thigh | ['tʃikin 'θai] | 鸡大腿 | starch | [staːtʃ] | 淀粉 |
| shank | [ʃæŋk] | 牛腱子 | scallion | ['skæliən] | 大葱 |
| lamb | [læm] | 羊腿 | watercress | ['wɔːtərkres] | 豆瓣菜 |
| spare ribs | ['spɛə 'ribz] | 排骨 | ginger | ['dʒindʒər] | 姜；生姜 |
| chicken breast | ['brest] | 鸡脯肉 | poultry | ['poultri] | 家禽 |
| cube | [kjuːb] | 四方块 | cater | ['keitə] | 满足需要 |
| chunk | [tʃʌŋk] | 不规则块 | season with ... | — | 以……调味 |
| dice | [dais] | 小丁 | soft fry | ['sɔft frai] | 软炸 |
| mince | [mins] | 碎末，肉馅；切碎，剁碎，绞碎 | quick-fry | ['kwik frai] | 爆炒 |
| block | [blɔk] | 大块 | mix...with | — | 混合…… |
| filet | [fi'lei] | 鱼片，肉片 | marinate | ['mærineit] | 腌，浸泡（食物） |
| roll | [rəul] | 卷 | crispy | ['krispi] | 脆的 |
| strip | [strip] | 条状 | | | |

## Learning Tips（学习提示）

把中餐菜名译成英文的时候，应尽量将菜肴的原料、烹制方法、菜肴的味型等翻译出来，让客人一目了然。现介绍 5 种翻译菜肴的方法。

1. 过去分词 + 菜肴主料：Stewed Spare Ribs（炖排骨）

2. 口味 + 菜肴主料：Crispy Chicken（脆皮鸡）

3. 过去分词 + 菜肴主料 +with+ 辅料：Fried Diced Pork with Green Peas（青豌豆肉丁）

4. 菜肴主料 +with+ 辅料或汤汁 Beef with Orange Peel（陈皮牛肉）

   Chicken with Ginger Oil（姜汁鸡）

5. 菜肴主料 + 形状 +with+ 辅料或汤汁 Beef Filet with Green Pepper（青椒牛柳）

   Fish Filet with Tomato Sauce（茄汁鱼片）

## Culture Knowledge（文化知识）

### 刀工的基本操作姿势

1. 两脚要自然站稳，与菜墩有适当的距离。上身略向前倾，前胸稍挺，不要弯腰弓背。两眼注视墩上两手操作的部位。

2. 右手握刀时，拇指与食指捏住刀箍，全手掌握好刀柄。左手控制原料，使原料平稳、不滑动，以便落刀。

3. 握刀时，手腕要灵活有力，菜墩的放置要适合自身的高低。

### 3 种肉类切法的要诀

1. 切肉块要诀：左手按住肉块，手指微弯顶住刀背，微斜式下刀切块，确认肉已切断后，再切下一块。记得要逆纹下刀，这样肉块才会有弹性。

2. 切肉片要诀：右手持刀，刀背倾向肉块那一侧，然后下刀，将片刀以前后拉锯法切出肉片。一定要用片刀之类的薄刀来切，肉片才会薄；用剁刀切片，成品较厚且不美观。

3. 切肉丝要诀：先用片刀将食材切成较厚的肉片，再切成肉丝，重点是刀要利且薄才切得断。此外，肉块建议先放到冰箱微微冷冻后，再切片或切丝，此时肉块会更容易切。

# Unit 9
# Seafood

## You will be able to:

⋄ get familiar with different kinds of seafood;
⋄ describe the seafood preparation process;
⋄ talk about the process of making seafood dishes.

# Warming Up (热身)

**1 Look and match** (看图并连线)

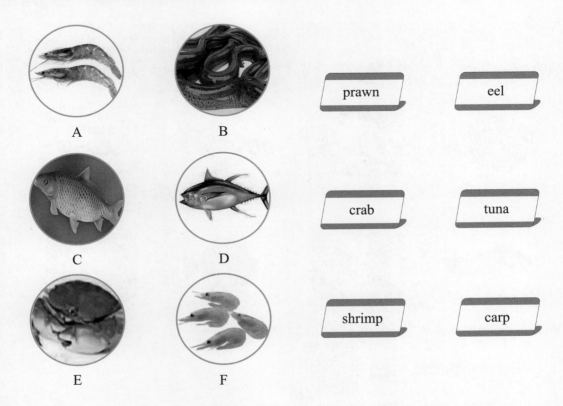

**2 Decide true or false** (判断正误)

1. _____ Eel is nutritious (有营养的), which fits for children and the old.

2. _____ Carp jumps over the dragon gate is a legend handed down in China for many years.

3. _____ Prawn is vegetable and shrimp is meat.

4. _____ Tuna can be made into the can.

5. _____ A crab has 8 legs, one head and two wings.

# Now I Can Learn（学一学）

## Part A

**1** Read and write（读单词和词组，写出鱼的英文表达）

| flatfish 比目鱼 | codfish 鳕鱼 | pomfret 鲳鱼 | ribbonfish 带鱼 |
| turbot 多宝鱼 | bass 鲈鱼 | grass carp 草鱼 | yellow croaker 黄花鱼 |

1._____ 2._____ 3._____ 4._____

5._____ 6._____ 7._____ 8._____

**2** Drill（句型训练）

**A** What do we call these fish tools in English?

**B** They are called fish scale and fish scissors.

**A** What shall I do with them?

**B** Scale the fish with the scale, gut the fish and take out the gills with fish scissors.

**3** Listen, complete and read（听录音，完成并朗读对话）

A  scale the fish with the scale

B  fish scale and fish scissors

C  Use the fish scissors to gut

D  get rid of the fish line

**Chef:** Today I'll teach you how to treat the grass carp.
**Jack:** Wonderful. What do we call these fish tools in English, chef?
**Chef:** They are called _____ .
**Jack:** What shall we do with them?
**Chef:** You can _____ and take out the gills with the scissors.
**Jack:** I've done it. And then?
**Chef:** _____ the grass carp.
**Jack:** Oh, these fish tools are very useful.
**Chef:** Yes. Don't forget to _____ and clean it.
**Jack:** I see.

## Part B

**1 Listen and read**（听录音，读单词和词组）

| | | | |
|---|---|---|---|
| lobster 龙虾 | oyster 牡蛎，生蚝 | clam 蛤蜊 | phoenix-tailed prawn 凤尾虾 |
| scallop 扇贝 | mussel 贻贝，青口 | abalone 鲍鱼 | sea whelk 海螺 |

**2 Look, write and talk**（看图写英文，并进行对话练习）

1. _____

2. _____

3. _____

4. _____

**A:** What shellfish would you like to eat?
**B:** I'd like to eat _____ .

8. _____

5. _____

6. _____

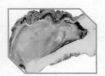
7. _____

## 3 Read（读重点词汇）

1. nip off the head　　　去掉头部
2. take out the vein　　 去掉泥肠
3. remind someone　　　 提醒某人
4. egg wash　　　　　　 蛋液

## 4 Choose and read（选择正确答案，并朗读对话）

**Jack:** What shall we do with the shrimp before frying, chef?

**Chef:** We should peel the shrimp and _____.

**Jack:** OK, it's so easy. And then?

**Chef:** _____.

**Jack:** Do you have anything to _____?

**Chef:** Make sure to _____ and dip in egg wash.

**Mike:** Thank you. I remember it.

A. Take out the vein

B. remind me

C. drain the shrimp off

D. nip off the head

## Part C

### 1 Read and tick（读单词和词组，勾出你喜欢吃的鱼产品）

☐ mandarin fish 鳜鱼　　☐ catfish 鲶鱼　　☐ silver fish 银鱼　　☐ silver carp 鲢鱼
☐ sturgeon 鲟鱼　　　　☐ trout 鳟鱼　　　☐ cuttlefish 墨鱼　　☐ crucian 鲫鱼

## 2 Match（连线）

1. clean the fish        A

2. cut open the fish     B

3. take out the gills    C

4. scale the fish        D

5. gut the fish          E

## 3 Fill in and evaluate（补全单词，并进行自我评价）

1. turb_t    2. p_mfret    3. b_ss    4. tr_ _t    5. _yster

6. l_bs_ _    7. scal_    8. shr_ _p    9. cl_m    10. gr_ss c_ _p

Assessment: If you can write 8–10 words, you are perfect.

If you can write 4–7 words, you are good.

If you can write 1–3 words, you need to try again.

Unit 9  Seafood  113

# Now I Can Speak（说一说）

## Make dialogues（对话练习）

e.g.
**A:** What shall I do with the *fish*?
**B:** *Scale it.*
**A:** With what?
**B:** *With a fish scale.*

shrimp / nip off the head / scissors
oyster / open / oyster knife
crucian / cut open / fish scissors

e.g.
**A:** What kind of seafood dish can you cook?
**B:** I can cook _____.
**A:** How do we call it in Chinese?
**B:** It is called _____.

| Fried Shrimp with Spicy Salt 椒盐炸虾 | Boiled Fish with Pickled Cabbage and Chili 酸菜鱼 | Steamed Mitten/Hairy Crabs 清蒸大闸蟹 | Braised Eel in Brown Sauce 红烧鳝鱼 |

# Now I Can Read（读一读）

**1** Read and answer（读对话，回答问题）

Jack: Could you tell me how to make the poached shrimp, chef?
Chef: No problem. First you should prepare the shrimps.
Jack: I know that. Cut off the mustache and get rid of the black line in its back. Is that right?
Chef: You are so clever. Then wash them well and put them into a basin.
Jack: OK, how about the ingredients?
Chef: Ginger slices, green onion sections, vinegar, salt, cooking wine and soy sauce.
Jack: I've finished them.
Chef: Let's do it now.

1. What shall we do with the black line?
   _____

2. What ingredients will they need?
   _____

**2** Write（参考上一题，用英文写出虾的加工过程）

1._____     2._____     3._____

## 3 Tick(勾出水煮大虾需要的调料)

- ☐ ginger  ☐ soy sauce  ☐ cooking wine  ☐ vinegar
- ☐ salt  ☐ sugar  ☐ green onion  ☐ meat

## 4 Read and decide(读菜谱,判读正误)

### Steamed Fish with Chopped Red Chili
### Ingredients

| | | |
|---|---|---|
| 1 pomfret | salt | 400 g green onion sections |
| ginger slices | vinegar | dry chili pepper |
| white sugar | soy sauce | five spices powder |
| cooking wine | | |

### Directions

1. Add in the cleaned pomfret. Turn to medium heat.
2. Fry the fish for about 1 minute on both sides.
3. Add the seasonings to heat over high heat until the juice is boiling.
4. Turn off the heat. Remove the pomfret from the wok and serve on a plate.

### Characteristics

Steamed Fish with Chopped Red Chili is salty and fresh with an attractive golden color.

1. _____ Steamed Fish with Chopped Red Chili is Guangdong dish.

2. _____ Steamed fish and poached fish have different tastes.

3. _____ Cooking wine is not necessary in this dish.

4. _____ We usually serve this dish in the mixing bowl.

5. _____ The fish will be fried on both sides.

## 5 Translate and write （菜品原料中英文互译，并写出菜品制作过程的英文表达）

### Ingredients

鲳鱼 _____     salt _____
青椒段 _____   white sugar _____
干辣椒 _____   vinegar _____
料酒 _____     sliced ginger _____

### Production process

**1** _____ the cleaned pomfret. Turn to _____ .

**2** Fry _____ .

**3**

**4**

## Now I Can Do （练一练）

**Exercise 1**   Translate into Chinese or English（单词中英文互译）

1. flatfish _____    2. 蛤蜊 _____

3. codfish _____    4. 扇贝 _____

5. ribbonfish _____        6. 鲍鱼 _____

7. pomfret _____          8. 海螺 _____

9. squid _____            10. 草鱼 _____

## Exercise 2  Tick off the odd words（勾出不同类别的单词或词组）

1. A. fish scissors    B. fish scale    C. cuttlefish    D. fish forceps
2. A. shrimp          B. mussel        C. salmon        D. oyster
3. A. take out        B. cut open      C. scale         D. scallop
4. A. delicious       B. terrible      C. gut           D. wonderful
5. A. vinegar         B. sugar         C. soy sauce     D. crab

## Exercise 3  Find the answers（找答案）

1. Could you tell me how to treat the shrimp before cooking?
2. How about the ingredients?
3. What is the best way to clean it?
4. How long shall I stew the fish?
5. What shall I do with the fish?

A. Scale it with the scale.
B. Sliced ginger, green pepper sections, sugar, vinegar and soy sauce.
C. Peel the body, nip off the head and pull out the vein.
D. Wash it with salted water.
E. 8 minutes.

## Exercise 4  Make sentences in right order（连词成句）

1. on, the, fish, fry, for, 1 minute, both sides, about

   _____

2. with, Steamed Fish with Chopped Red Chili, attractive, an, golden color, salty, is, and, fresh

   _____

3. well, wash, put, into, the, a basin, shrimps, and

   _____

4. nip off, and, take, out, the head, the vein

   _____

5. what, shellfish, you, eat, would, of, kind, like to

   _____

## Exercise 5    Write（写出下列菜品的英文表达）

> West Lake Fish in Vinegar Gravy      Steamed Wuchang Fish
> Braised Ribbonfish in Brown Sauce    Sautéed Shrimp with Broccoli
> Sautéed Crab in Hot Spicy Sauce      Sweet and Sour Mandarin Fish

松鼠鳜鱼_____        香辣蟹_____
红烧带鱼_____        西湖醋鱼_____
清蒸武昌鱼_____        翡翠虾仁_____

## Exercise 6    Translate into Chinese or English（句子中英文互译）

1. 不同的工具有不同的用途。

2. 请用这把特别的剪刀剪开鱼腹。

3. 麦克和师傅正在做白灼基围虾。

4. Add the seasonings to heat over high heat until the juice is boiling.

5. Remove the pomfret from the wok and serve on a plate.

**Word List**(单词表)

| | | | | | |
|---|---|---|---|---|---|
| prawn | [prɔːn] | 对虾 | scallop | [skɔləp] | 扇贝 |
| crab | [kræb] | 螃蟹 | abalone | [æbəˈləuni] | 鲍鱼 |
| shrimp | [ʃrimp] | 虾 | sea **whelk** | [welk] | 海螺 |
| eel | [iːl] | 鳝鱼 | **mandarin** fish | [ˈmændərin] | 鳜鱼 |
| tuna | [ˈtjuːnə] | 金枪鱼 | catfish | [ˈkætfiʃ] | 鲶鱼 |
| carp | [kɑːp] | 鲤鱼 | **silver** fish | [ˈsilvə(r)] | 银鱼 |
| flatfish | [ˈflætfiʃ] | 比目鱼 | **silver** carp | [ˈsilvə(r)] | 鲢鱼 |
| codfish | [kɔdfiʃ] | 鳕鱼 | sturgeon | [ˈstəːdʒən] | 鲟鱼 |
| pomfret | [ˈpɔmfret] | 鲳鱼 | trout | [traut] | 鳟鱼 |
| ribbonfish | [ˈribənfiʃ] | 带鱼 | cuttlefish | [ˈkʌtlfiʃ] | 墨鱼 |
| turbot | [ˈtəːbət] | 多宝鱼 | crucian | [ˈkruːʃən] | 鲫鱼 |
| **grass** carp | [grɑːs] | 草鱼 | gut | [gʌt] | 取出内脏 |
| yellow **croaker** | [ˈkrəukə] | 黄花鱼 | gill | [gil] | 鱼鳃 |
| lobster | [ˈlɔbstə(r)] | 龙虾 | **nip** off the head | [nip] | 去掉头部 |
| oyster | [ˈɔistə(r)] | 牡蛎，生蚝 | take out the **vein** | [vein] | 去掉泥肠 |
| clam | [klæm] | 蛤蜊 | **remind** someone | [riˈmaind] | 提醒某人 |
| **phoenix**-tail prawn | [ˈfiːniks] | 凤尾虾 | egg wash | — | 蛋液 |

# Learning Tips（学习提示）

海鲜一向是受人们欢迎的食物。海鲜含有丰富的蛋白质、低胆固醇以及各种微量元素。与其他肉类相比，海鲜更有益于人的健康。

但是，食用海鲜的禁忌，你知道吗？

1. 海鲜煮不熟含有细菌
2. 死贝类病菌毒素多
3. 海鲜与啤酒同吃惹痛风
4. 海鲜与水果同吃会腹痛
5. 吃海鲜后喝茶长结石
6. 冰鲜虾不可白灼着吃
7. 海鲜与维C同食会中毒
8. 关节炎患者忌多吃海鲜

# Culture Knowledge（文化知识）

西湖醋鱼又名叔嫂醋鱼。相传南宋时期有宋姓兄弟隐居于西湖，享受着打鱼之乐。宋兄有一妻，美艳动人，偶有一日，宋妻在湖边浣衣被恶霸看上，横生霸占之意，便用阴毒之法害死宋兄。叔嫂二人一起上官府告状，但官府与恶霸串通一气，此事便不了了之。叔嫂二人惧怕恶霸的打击报复，便决定分别离开此地。临走之前，宋嫂为小叔烧制了一道隐含深刻意义的鱼，加糖加醋，寓意人生又酸又甜，既不能忘记哥哥的死，也不能因为生活得好而忘记人间疾苦。后来，宋弟考取了功名，衣锦还乡，为兄长报了仇，可是却找不到对自己人生有所指点的嫂嫂了。后宋弟参加一次官宴，在宴席上又吃到了熟悉的味道，便去寻烧制此鱼的人，结果此人正是为了躲避恶霸报复隐姓埋名、在官厨里当下手的嫂嫂。叔嫂再次相遇，欣喜至极……

西湖醋鱼在不同时期的做法是不同的，清代袁牧在《随园食单》中所记：鱼切大块，油煎后放酱、醋、酒等，烧熟迅速起锅。而在清末民初的文献记录中，醋鱼有一鱼三吃的说法：一半做鱼生拌胡椒面和麻油食用，一半做醋鱼，鱼骨则用来熬汤。在梁实秋先生《雅舍谈吃》中的醋鱼则是完全汆水而吃，所用的糖醋酱没有当今的酱黑色，而是少用酱油提高糖醋酱的透明度。

# Unit 10
# Staple Food & Snacks

*You will be able to:*

✧ get familiar with some Chinese staple food and snacks;
✧ know the basic sentences about staple food and snacks in English;
✧ talk about the process of making simple staple food.

# Warming Up(热身)

**1** Look and match(看图并连线)

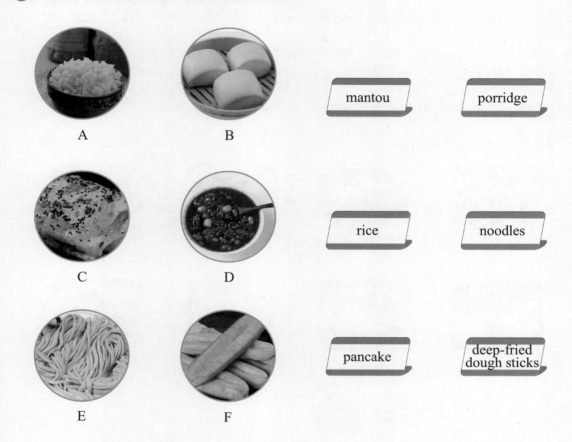

**2** Decide true or false(判断正误)

1. _____ Noodles are the main food in the southern Chinese diet.

2. _____ Porridge is a common breakfast in Beijing.

3. _____ Noodles are made from wheat flour.

4. _____ Rice is a main food for people in southern China.

5. _____ Deep-fried dough sticks are a kind of steamed food.

# Now I Can Learn（学一学）

## Part A

### 1 Read and write（读单词和词组，写出下列面食的英文表达）

| | |
|---|---|
| beef noodles 牛肉面 | stir-fried Shanxi pasta 山西烩猫耳朵 |
| plain noodles 阳春面 | hot dry noodles 热干面 |
| noodles with soy bean paste 炸酱面 | dandan noodles 担担面 |
| Henan braised noodles 河南烩面 | noodles with wonton 云吞面 |

1._____  2._____  3._____  4._____

5._____  6._____  7._____  8._____

### 2 Drill（句型练习）

**A** What do you recommend?

**B** Beef noodles, dandan noodles, and noodles with soy bean paste.

**A** What is the eating habit in China?

**B** In the north of China, people eat lots of noodles and mantou. In the south of China, people eat a lot of rice and seafood.

### 3 Listen, complete and read（听录音，完成并朗读对话）

**A** people eat lots of noodles and mantou

**B** if I introduce some famous food

**C** It is mainly divided into two parts

**D** people eat a lot of rice

**Jack:** What is the eating habit in China?

**Chef:** As we all know, China is large in its size. _____ by Changjiang River. In the north of China, _____ . In the south of China, _____ .

**Jack:** What do you recommend _____ to foreigners?

**Chef:** Beef noodles in Lanzhou, dandan noodles in Sichuan, noodles with soy bean paste in Beijing, plain noodles in Shanghai and so on.

**Jack:** Wow, sounds great.

**Chef:** Don't forget to have a try when you have time.

## Part B

### 1 Listen and read（听录音，读单词和词组）

steamed rice rolls 肠粉　　　cha siu bao 叉烧包　　　shaomai 烧卖
steamed glutinous rice chicken 糯米鸡　　　Malaysian sponge cake 马来糕
egg tart 蛋挞　　　steamed shrimp dumpling 虾饺　　　steamed creamy custard bun 奶黄包

### 2 Look, write and talk（看图写英文，并进行对话练习）

1. _____

2. _____

3. _____

4. _____

A: What would you like to eat for breakfast?
B: I'd like to eat _____ .

8. _____

5. _____

6. _____

7. _____

## 3 Read（读重点词汇）

1. prepare the filling 准备馅
2. make a dough 和面
3. roll the dough into wrappers 擀饺子皮
4. pack the dumplings 包饺子

## 4 Choose and read（选择正确答案，并朗读对话）

**Jack:** Could you tell me how to make steamed shrimp dumplings?
**Chef:** No problem. First, we need to _____ .
**Jack:** And then?
**Chef:** In the meantime, we will _____ .
**Jack:** Right now?
**Chef:** Yes. Now you need to _____ .
**Jack:** I see. It is not difficult.
**Chef:** Wonderful. Then put some filling in the middle of the round wrapper and _____ .
**Jack:** And after that?
**Chef:** Steam the dumplings for 10 to 15 minutes, and then it can be served.

A. roll the dough into wrappers
B. pack the dumplings
C. make a dough
D. prepare the filling

## Part C

### 1 Read and tick（读单词和词组，勾出你喜欢吃的中国传统食物）

☐ dumpling 饺子　　☐ zongzi 粽子　　☐ tangyuan 汤圆
☐ mooncake 月饼　　☐ spring rolls 春卷　　☐ niangao 年糕

## 2 Match（连线）

1. prepare the filling     A

2. roll the wrappers     B

3. pack the dumplings     C

4. make a dough     D

5. boil the dumplings     E

## 3 Fill in and evaluate（补全单词，并进行自我评价）

1. wr__pper    2. f__lling    3. st____med    4. d____pling    5. d__ugh

6. p__ck    7. prepar__    8. r__ll    9. n____dle    10. p__rri__ge

Assessment: If you can write 8–10 words, you are perfect.

            If you can write 4–7 words, you are good.

            If you can write 1–3 words, you need to try again.

# Now I Can Speak (说一说)

## Make dialogues (对话练习)

**e.g.**
**A:** What shall I do with the *dough*?
**B:** *Roll it*.
**A:** With what?
**B:** With *the rolling pin*.

dumpling dough / knead it / hands
dumpling filling / stir it / chopsticks
shrimp dumplings / steam them / steamers

**e.g.**
**A:** What kind of food can you cook?
**B:** I can cook _____.
**A:** What do we call it in Chinese?
**B:** It is called _____.

chowmein 炒面   sliced noodles 刀削面   beef noodles 牛肉面   Yangzhou fried rice 扬州炒饭

# *Now I Can Read*（读一读）

**1** Read and answer（读对话，回答问题）

**Jack:** Could you tell me how to make vegetable chowmein?
**Chef:** No problem. First you should prepare the noodles.
**Jack:** I see. Boil the noodles, is that right?
**Chef:** You are so clever. When the noodles are soft, drain them and set aside.
**Jack:** OK, how about the vegetables?
**Chef:** Slice the red bell peppers and onions, cut the garlic sprouts into sections. Then stir-fry them with some ginger and garlic.
**Jack:** I've finished them. And after that?
**Chef:** Add some salt and chicken essence in the noodles, and then mix well.

1. What kind of noodles will they cook?
   _____

2. What ingredients will they need?
   _____

**2** Write（参考上一题，用英文写出蔬菜的加工过程）

1. _____    2. _____    3. _____

Unit 10  Staple Food & Snacks

### 3 Tick（勾出蔬菜炒面需要的调料和配料）

☐ ginger  ☐ garlic  ☐ red bell pepper  ☐ vinegar
☐ salt    ☐ sugar   ☐ onion            ☐ meat

### 4 Read and decide（读菜谱，判读正误）

**Yangzhou Fried Rice**（扬州炒饭）

**Ingredients**

1 egg
20 g boiled shelled shrimps
20 g boiled green beans
chopped scallion
50 g diced sausages
50 g diced carrots
1 bowl cooked rice
salt and white pepper

**Directions**

1. Heat oil in the wok. Drop in the eggs and scramble. Then set aside.
2. Add in the chopped scallions, diced carrots, boiled green beans, diced sausages and boiled shrimps, and stir-fry them.
3. Add in cooked rice. Stir all together. Then pour in the eggs and mix them.
4. Season with salt and white pepper to taste.

**Tips**

It is better to use refrigerated rice for fried rice as it does not stick together.

1. _____ Yangzhou fried rice is Beijing dish.

2. _____ Steamed rice and fried rice are different cooking ways.

3. _____ Sugar is not necessary in this dish.

4. _____ Refrigerated rice will make your fried rice perfect.

**5** **Translate and write**（菜品原料中英文互译，并写出扬州炒饭制作过程的英文表达）

## Ingredients

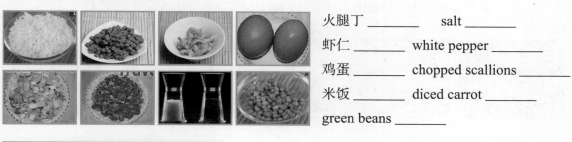

火腿丁 _____     salt _____
虾仁 _____     white pepper _____
鸡蛋 _____     chopped scallions _____
米饭 _____     diced carrot _____
green beans _____

## Production process

1 _____

2 _____

3 _____

4 _____

# Now I Can Do（练一练）

**Exercise 1**  Translate into Chinese or English（单词和词组中英文互译）

1. guotie _____     2. 蛋挞 _____
3. spring rolls _____     4. 饺子 _____
5. beef noodles _____     6. 炒面 _____

7. mooncake _____          8. 云吞面 _____

9. steamed rice rolls _____          10. 烧卖 _____

## Exercise 2   Look and match（将中国传统节日和饮食习俗连线）

 ①  ②  ③  ④  ⑤

a. mooncake    b. tangyuan    c. jiaozi    d. zongzi    e. spring rolls

## Exercise 3   Choose the best answers（选择正确答案）

1. Place the dough on a floured board and roll it _____ .
   A. on            B. out           C. up            D. of
2. At last, the chef will sprinkle the scallions _____ top.
   A. up            B. on            C. in            D. above
3. Please steam shrimp dumplings _____ 10–15 minutes.
   A. on            B. for           C. of            D. out
4. Divide the mixture into walnut size balls after _____ well.
   A. is kneading   B. kneading      C. knead         D. to knead
5. Onions and spices may be _____ to make your rice savory.
   A. add           B. adding        C. added         D. to add
6. What kind of noodles shall I _____ ?
   A. make          B. cut           C. study         D. look
7. Can you tell me how to _____ scrambled eggs?
   A. making        B. made          C. make          D. be made
8. First, let's heat the oil _____ the wok.
   A. on            B. in            C. of            D. to
9. Pour _____ the eggs, mix them, season _____ salt to taste.
   A. in... of      B. on... to      C. in... with    D. to... with
10. Cover the pan and _____ the heat to very low.
    A. place         B. reduce        C. break         D. mix

132  中餐烹饪英语

## Exercise 4  Write（写出下列主食的英文表达）

> sliced noodles
> steamed twisted roll
> fried baked scallion pancake
> cold noodles with seasame sauce
> fried rice with seafood
> wotou with black rice

凉面_____  葱油饼_____
刀削面_____  海皇炒饭_____
蒸花卷_____  黑米小窝头_____

## Exercise 5  Translate into Chinese or English（句子中英文互译）

1. 蛋炒饭做好了。
2. 把拌好的馅用饺子皮包好。
3. 我们用西红柿和鸡蛋做什么？
4. Add the green peas, diced carrots, salt and stir well.
5. Please crack three eggs in the bowl.

 # Word List（单词表）

| | | | | | | |
|---|---|---|---|---|---|---|
| mantou | — | 馒头 | steamed glutinous rice chicken | [stiːmd 'gluːtnəs rais 'tʃikin] | 糯米鸡 |
| rice | [rais] | 米饭 | Malaysian sponge cake | [mə'leiʒn spʌndʒ keik] | 马来糕 |
| porridge | ['pɔridʒ] | 粥 | steamed shrimp dumpling | [stiːmd ʃrimp dʌmpliŋ] | 虾饺 |
| noodles | [nuːdlz] | 面条 | steamed creamy custard bun | ['stiːmd 'kriːmi 'kʌstəd bʌn] | 奶黄包 |
| pancake | ['pænkeik] | 煎饼 | dumpling | ['dʌmpliŋ] | 饺子 |
| deep-fried dough sticks | [diːp fraid dəu stiks] | 油条 | zongzi | — | 粽子 |
| beef noodles | [biːf 'nuːdlz] | 牛肉面 | tangyuan | — | 汤圆 |

Unit 10  Staple Food & Snacks

续表

| | | | | | |
|---|---|---|---|---|---|
| stir-fried Shanxi pasta | [stə: fraid shanxi 'pæstə] | 山西烩猫耳朵 | mooncake | ['mu:nkeik] | 月饼 |
| plain noodles | [plein 'nu:dlz] | 阳春面 | spring rolls | ['spriŋ rəulz] | 春卷 |
| hot dry noodles | [hɔt drai 'nu:dlz] | 热干面 | niangao | — | 年糕 |
| noodles with soy bean paste | ['nu:dlz wið sɔi bi:n peist] | 炸酱面 | chowmein | — | 炒面 |
| dandan noodles | [dandan 'nu:dlz] | 担担面 | sliced noodles | [slaist 'nu:dlz] | 刀削面 |
| Henan braised noodles | [henan breizd 'nu:dlz] | 河南烩面 | Yangzhou fried rice | [yangzhou fraid rais] | 扬州炒饭 |
| noodles with wonton | ['nu:dlz wið wɔn 'tɔn] | 云吞面 | chopped scallions | [tʃɔpt 'skæliənz] | 葱花 |
| steamed rice rolls | [sti:md rais rəulz] | 肠粉 | red bell pepper | [red bel 'pepə] | 红柿子椒 |
| cha siu bao | — | 叉烧包 | make a dough | — | 和面 |
| shaomai | — | 烧卖 | prepare the filling | — | 准备饺子馅 |
| egg tart | [eg tɑ:t] | 蛋挞 | roll the wrappers | — | 擀饺子皮 |

# Learning Tips（学习提示）

北京，我国的首都，也是一座古都。北京小吃博采四方小吃之精华，兼收各族小吃之特色。北京小吃现逾百种，形成了蒸、煮、煎、炸、烤、烙、爆、冲等多种技艺，其间融合多民族的传统食艺、食俗，形成了琳琅满目、缤纷斑斓的诱人品相。受人们喜爱的北京小吃有豆面糕、艾窝窝（ai wowo）、糖卷果、姜丝排叉、糖耳朵（fried sugar cake）、面茶、馓子麻花、萨其玛、焦圈（fried ring）、糖火烧、豌豆黄（pea flour cake）、豆馅烧饼、奶油炸糕（fried creamy cake）等，号称"老北京十三绝"。

# Culture Knowledge（文化知识）

中式面点品种繁多，其主要分类方法有以下几种。

**按地理区域分类**：中式面点大体上可以划分为南味和北味两大类型。北味以面粉、杂粮制品为主；南味以米、米粉制品为主。现在一般以京鲁风味（简称京式）为北味的代表。江苏一带面点（简称苏式）花色繁多、做工精细、味道偏甜，被普遍认为是南味的主流之一；广东一带的面点（简称广式）较多地吸收了西式面点的制作方法，体现了南国风味的制作特色，是南味面点的后起之秀。

**按使用原料分类**：根据面点的制作原料，人们习惯将其分为麦类制品、米类制品、杂粮类制品和其他制品。**按所用馅料分类**：根据制品有无馅料可分为有馅制品和无馅制品两种。有馅制品又可分为荤馅类、素馅类和荤素混合馅类3种。**按制品口味分类**：可分为甜点、咸点、甜咸味和无味类制品4类。

**按熟制方法分类**：可分为煮制品、蒸制品、炸制品、烤制品、煎制品、烙制品及复合熟制品。**按制品形态分类**：可分为糕类、团类、饼类、饺类、条类、粉类、包类、卷类、饭类、粥类、冻类、羹类等。**按成品干湿分类**：可分为干点制品、湿点制品和水点制品等。在中国人民的饮食生活中，中式面点占有相当重要的地位。

Unit 10　Staple Food & Snacks　135

# 参考文献

［1］赵丽.烹饪英语［M］.北京：北京大学出版社，2008.

［2］宋洁.烹饪英语［M］.2版.北京：中国轻工业出版社，2020.

［3］孙诚.烹饪英语［M］.北京：高等教育出版社，2009.

［4］杜纲.烹饪英语［M］.2版.重庆：重庆大学出版社，2022.

［5］张毅，贾颖丽.烹饪厨房英语［M］.2版.重庆：重庆大学出版社，2021.